DRY WATER

DRY WATER
DIVING HEADFIRST INTO AFRICA
TAMMIE MATSON

HODDER

*For my family in Australia and my surrogate families in Africa —
your support over the years means the world to me.*

HODDER AUSTRALIA

Published in Australia and New Zealand in 2006
by Hodder Australia
(An imprint of Hachette Livre Australia Pty Limited)
Level 17, 207 Kent Street, Sydney NSW 2000
Website: www.hachette.com.au

Reprinted 2006

Copyright © Tammie Matson 2006

This book is copyright. Apart from any fair dealing
for the purposes of private study, research, criticism
or review permitted under the *Copyright Act 1968*,
no part may be stored or reproduced by any process
without prior written permission. Enquiries should
be made to the publisher.

**National Library of Australia
Cataloguing-in-Publication data**

Matson, Tammie K. 1977–
  Dry water : diving head first into Africa.

  ISBN 0 7336 1984 3.

  1. Matson, Tammie K. 1977– . 2. Wilderness conservationists -
  Australia - Biography. 3. Wildlife conservation - Africa.
  4. Wildlife management - Africa. I. Title.

591.092

Cover design by Christabella Designs
Cover photography by Peter and Beverley Pickford
Maps and photo sections by Kinart
Internal photography by Tammie Matson unless otherwise credited
Text design and typesetting by Bookhouse, Sydney
Printed in Australia by Griffin Press, Adelaide

Hachette Livre Australia's policy is to use papers
that are natural, renewable and recyclable products
and made from wood grown in sustainable forests.
The logging and manufacturing processes are expected
to conform to the environmental regulations
of the country of origin.

CONTENTS

Prologue *1*

A Great Leap into the Unknown *8*
The Deep End *17*
This is Zambia *37*
Where the Wild Things Are *57*
The Hippo Lady and the Wild Dog Man *76*
An Impala Initiation *94*
Serendipity *115*
Out of Disaster and into the Desert *129*
What it Feels Like to Be Prey *146*
Suicide Season *165*
One of the Boys *181*
Survival of the Fittest *196*
Walking with the Impalas *215*
Forces of Nature *231*
An Ending and a Beginning *248*

SOUTHERN AFRICA

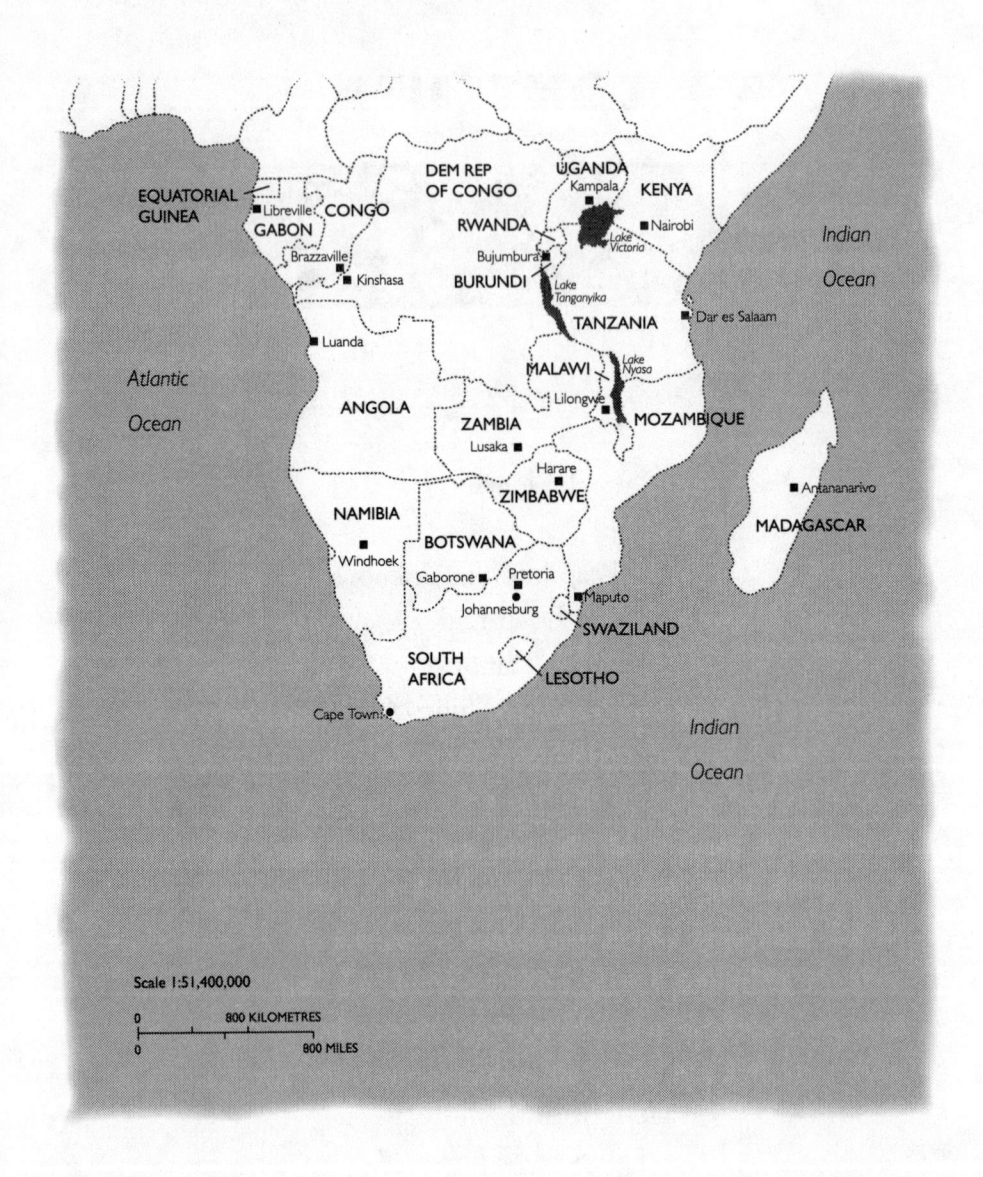

NAMIBIA

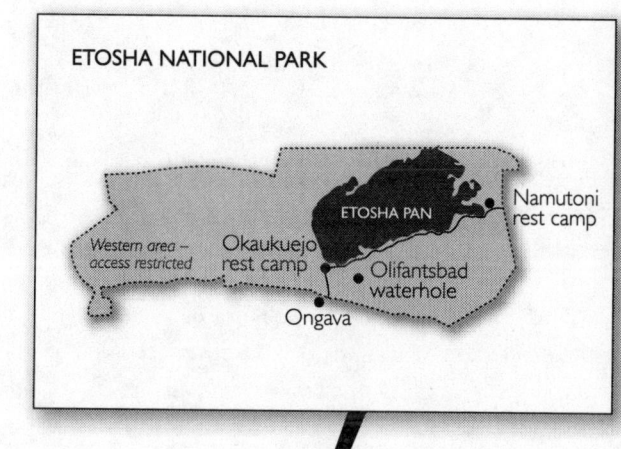

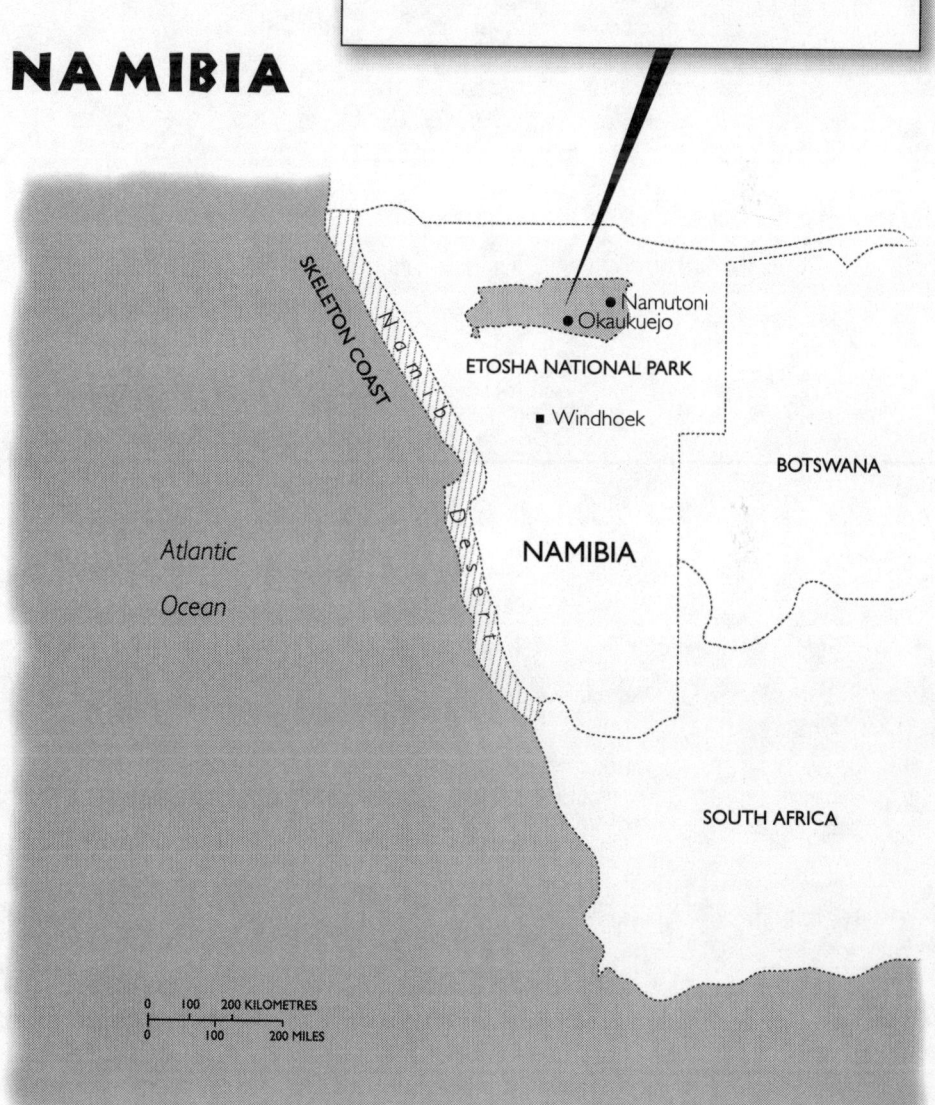

ETOSHA NATIONAL PARK

- Western area – access restricted
- Okaukuejo rest camp
- ETOSHA PAN
- Namutoni rest camp
- Olifantsbad waterhole
- Ongava

- Namutoni
- Okaukuejo
- ETOSHA NATIONAL PARK
- Windhoek

SKELETON COAST

Namib Desert

Atlantic Ocean

NAMIBIA

BOTSWANA

SOUTH AFRICA

0 100 200 KILOMETRES
0 100 200 MILES

PROLOGUE

Something was dead. If I'd been back home in Australia I'd have checked the side of the road for a recently heaven-sent kangaroo. The rotting meat stunk like yesterday's road kill. As always, the stench of death stopped me in my tracks. That smell never ceased to fill me with an acute awareness of the gory side of life in the African bush. Namibia, my adopted home, had a way of always reminding me that in nature, life and death dance together in the same seductive rhythm. The sense that I was constantly dancing on the brink of the unknown textured my life with challenges and adventures. Maybe that was why I loved it there.

Because invariably, in Africa, life *and* death were all around me. Today was no different. I'd been driving west along the wide gravel track that forms the main road across the vast Etosha National Park, only vaguely aware of the heat haze and dust-blurred herds of springbok and zebra I passed. I'd been thinking instead of the cute guy I had a crush on, wondering when I would see him next. It could be any time between a day, three months or never again. Life in Etosha didn't offer much prospect of a love life. Where romance was concerned, it was a desert land and I was in the proverbial drought. Fortunately, I wasn't alone.

My friend, research volunteer and fellow single gal, Naomi, piped up in the passenger seat of my single-cab *bakkie* and announced with interest, 'Hey, there's a carcass over there near the waterhole. It looks like an elephant.'

I drove into the parking area near the Ozonjuitji m'Bari waterhole and switched off the engine. It was the first time either of us had seen a mountain of death so large, a mound of grey skin folded around swelling cavities and decomposing meat. I agreed with Naomi. Through my binoculars it looked like the body of a young elephant.

It was difficult to know how old the carcass was. From where we sat in the car, fifty metres away, the occasional oven-hot gusts of wind in our direction brought with them a rancid stench – testimony to the fact that the animal had been dead for a couple of days at least.

Hundreds of animals had converged on the waterhole, having trekked for kilometres for the blessing of water. They gathered like sinners at the confessional, desperate for salvation, for redemption from thirst. It was midmorning and already warming up to the forty-degree Celsius mark. Zebras whinnied, kicked and bit each other viciously, tearing into shadow-striped hides and drawing fresh blood, frustrated by the waiting and the interminable thirst. Oryx butted the odd courageous springbok that tried to sneak down to the waterline, their spear-like horns forming lethal weapons. Still more animals were emerging from the bush, forming an endless stream. Dust swirled around the whole congregation, like smoke at the gates of hell.

It was the end of November and it hadn't rained for at least six months in Etosha. Cicadas whirred deafeningly; maniacal, feverish, maddening. The mopane trees that dominated this environment

PROLOGUE

were beginning to sprout their butterfly-shaped leaves at last, after months as skeletons, creating a refreshing greenness in the bush again and providing valuable shade. Their new leaves, still pink and soft, created the illusion of abundance. New mopane leaves could be consumed by some of the browsers, but by the time the leaves were fully grown, toxins and cellulose made them indigestible to most species. Etosha, I was learning, was a place of illusions.

Not yet painted with the fresh white dung of vultures, and with its backside to us, I couldn't tell what had been feeding on the elephant carcass. Something clearly had. There was a gaping hole where the animal's soft-skinned stomach used to be.

Perhaps it had died of anthrax, the endemic disease that claims the lives of hundreds of animals every year in Etosha. Anthrax thrives in the man-made gravel pits that were dug to make Etosha's roads, but this particular strain is a far cry from the type sent by terrorists and designed to kill people. Animals ingest the anthrax bacterium (*Bacillus anthracis*) by eating infected grass or carcasses and inhaling contaminated dust. Elephants may get the disease when they throw dust over themselves as protection against heat and insects. In other parts of southern Africa, elephants are culled regularly to keep the numbers manageable. In contrast, Etosha's elephants naturally drop like flies from anthrax at the end of the dry season. As a result, the disease is seen as a gory but natural form of population control.

We had been instructed to use gloves to check carcasses for infection, and I always carried a testing kit in the car. It was simply a matter of taking a blood sample. As a resident PhD student based in Etosha, I was happy to help with routine park monitoring activities. Without much additional thought, clutching the anthrax kit under my arm, I headed over to check the elephant out, with Naomi following close behind.

Naively, we began walking towards the mound of rotting meat. Both of us had our eyes on the carcass, hypnotised by its putrefaction. As I got closer I could see that writhing maggots had invaded the remains of the stomach cavity. We were about thirty metres from the carcass and about the equivalent distance from the car. The smell was sickening and grew more pungent the closer we got. I turned my face away in the direction of the waterhole and covered my nose.

As I did so, I noticed that there was an opening in the dense wall of game at the waterhole. The animals parted like two curtains, seemingly in slow motion, forming an undulating wave of hooves and hide. Through the gap stalked three sultry lionesses.

Was this another illusion? No, I thought sickeningly. This was real. Their shoulders hunched low in hunting mode, the sinuous forms were already walking rapidly in our direction, clearly focused on two silly girls and what I now realised was their kill. The mass of animals at the waterline had formed a brilliant camouflage for them. Why hadn't I realised that the number of animals had been building up but not abating? None had left the waterhole. None had been drinking. They had just been standing. Agitated. On edge. Now I understood why. The lionesses, ingenious predatory queens, had been sitting at the water's edge, lazily watching, waiting for an animal to move too close, to grow too desperate for water to care about the consequences. And now they had found their prey in two foolish women and we were at least thirty metres from the car, our refuge.

Instantly my heart pounded so loudly that my ears hurt. Adrenaline surged through me like a lightning bolt before I had a chance to fully register what was going on and my limbs spun around in an ungainly pirouette.

'Fuck!' I exclaimed urgently, 'there's lions!'

PROLOGUE

I barely registered the look of disbelief and then alarm on Naomi's face as she came to an abrupt halt. There was no time for explanations. She had seen the lions now and knew we were in trouble. A second too long spent digesting the information could be fatal – and could lead to *us* being digested. We had to get back to the car, but we couldn't run, lest we provoke a full-on charge from the lions. There is nothing more appealing to a predator than a running target.

My adrenaline-fuelled legs carried my body hurriedly back in the direction of the car. Step by anguished step. Slow... Slow! Every instinct screaming at me to sprint, but logic forced me to walk, albeit quickly. Would we make the distance? I couldn't tell. They had a head start on us. I could hear Naomi's footsteps hard on my heels. My own sharp intakes of breath echoed in my ears. I had one eye on the car and one on the lionesses, weighing up the distances.

We were close now, about ten metres from safety. But the lionesses had gained on us, and they were no longer walking but trotting, coming right for us. I could feel their determined yellow eyes upon my back as I lengthened my stride, each step seemingly in slow motion, while the predators were in fast-forward mode. I could almost feel the agony of sharp teeth around my throat, flesh-tearing claws upon my back, the acrid smell of feline breath in my face.

Just when I thought we wouldn't make it, I reached the car and fell into it, crumpling in a terrified heap. Naomi reached the passenger seat a second later. My chest was heaving and adrenaline continued to surge through me, as if my body hadn't realised that I was safe. My hands were shaking and I forced them to stay still by gripping the steering wheel hard. I rested my forehead on the wheel, replaying the events in my head. A reverent silence embraced us both. And we stayed like that, in profound relief, for several minutes.

'Man, that was close,' I said finally.

Naomi and I looked at each other and broke into wide grins, then suddenly erupted with laughter that began somewhere deep in our bellies and graduated into torrents of great, heaving spasms that sent tears running down our faces. We felt stupid to have taken such a blind risk, but so grateful that we had been delivered unharmed to live another day under African skies.

Rational thought began to return. I wiped the tears of relief from my cheeks.

'You know, I don't think lions hunt in the day. Too hot for them or something,' I said.

Naomi shrugged. 'To be honest, I don't think we were in any real danger. They were probably just protecting their carcass and saw us as competition.'

I radioed Park headquarters to let them know that we'd found the carcass of a dead elephant at m'Bari but that we couldn't get close to it to take an anthrax sample because of a little workplace hazard.

'Tammie, are you sure you're looking at an elephant?' Wilfred's disgruntled voice echoed over the airwaves. 'There's a rhino carcass there that I'm aware of. We got an anthrax sample yesterday.'

My stomach dropped. As if it wasn't stupid enough of me to walk up to a carcass without having a good look around, I'd also managed to identify it incorrectly! I couldn't have felt more foolish. On the food chain of the scientists at Etosha, I was right down at the bottom with the worms and the dung beetles. And man, today I knew it.

By now the lionesses had slunk back to the rhino carcass and were guarding it protectively. Lions don't hunt in the daytime heat – I'd read that somewhere, I was sure of it. It made sense that a creature that slept for twenty hours a day wouldn't want to overdo

PROLOGUE

it when the sun was at its hottest. What's more, there was a carcass there with loads of meat in it – why would they need to hunt us? They were not going for us at all; they just wanted us to get away from the carcass. I laughed, thinking what an idiot I'd been to think that we were really in danger.

As I was about to turn the key in the ignition and rev the ute into action one of the lionesses began stalking a springbok ewe. The delicate antelope's fate was sealed in an instant by the suffocating grip of the lioness's mouth around her throat. The minute she made the kill, just metres from the car, the other lionesses loped over eagerly to grab their share of the spoils. I knew that Naomi was thinking the same thing I was: that springbok could have been us.

The lionesses had barely touched the springbok when out of the grass emerged a large black-maned male. He'd been there all along, invisible to us until now. His movements were deliberate as he approached the kill, his authority undeniable, palpable. All it took was a vicious snarl from him for the lionesses to relent. They made little objection apart from an annoyed snarl or two. The male picked up the springbok in his mouth, its entrails hanging in the dirt, and dragged it into the long, custard-coloured grass where he could eat alone, away from the scowls of the huntresses.

It was ominously clear to me and Naomi that our guardian angels had been on duty today. I realised then that in the African bush, without horns or claws to defend ourselves, we supposedly evolutionarily advanced humans are nothing more than prey.

A GREAT LEAP INTO THE UNKNOWN

Some of the best things in life happen by accident. My first trip to Africa was like that. I have my mother's fear to thank for it. If Mum hadn't been too scared to join my father on his lifelong dream to hunt in Africa, I never would have gone. If I hadn't been the eldest child in the family and, at the age of fifteen, considered old enough to appreciate the experience, perhaps Dad never would have taken me. Accidents and coincidences may have led me, an ordinary Australian girl, to Africa, but it has been sheer determination and an inexplicable love for the place that has made Africa a major part of my life for the past twelve years.

People often ask me why I keep going back and why I choose to live so far away from the relatively luxurious and easy life in Australia. I never have a satisfactory answer. All I can say is that Africa and I must be old friends, because no matter how much time I spend away, when I return to the warm coals of its campfires and feel the mellow glow of its sunsets, a sense of fond familiarity washes over me and it is as though we have never parted.

A GREAT LEAP INTO THE UNKNOWN

Africa is a complex land where inspiration and natural beauty mingle freely with suffering and unspeakable pain. To survive in this continent, Africans know that you have to be adaptable and willing to embrace change, because the wind can change direction at any moment and turn a metaphorical placid rhino into a blast of charging fury. Nothing is stable. Everything is always evolving. In life, sometimes you have to be able to see things from a different angle or as someone else perceives them. In Africa, it's an absolute necessity. That's the first thing I learned: to embrace a change in perspective.

***

It was the second day of my father's plains-game trophy hunt at Humani Ranch in Zimbabwe. It was June, the winter of 1993. I was almost sixteen and really didn't understand my father's desire to hunt at all. We were both bleary-eyed with jetlag and the last thing Dad or I expected was for him to shoot an animal so soon after we had arrived. When the opportunity presented itself – an old male wildebeest grazing among a herd of graceful impala – one of the ranch's professional hunters, Cuan, instructed Dad to grab his rifle.

'Tammie,' Cuan ordered matter-of-factly, 'you stay here with the guys. They'll look after you.'

Cuan's excitement was clear in his striking blue eyes. My first instinct was to feel put out at being left behind, but Dad explained later that the fewer people stalking, the better his chance of getting up close to his target. John and Pornai, Cuan's Shona trackers, grinned in a way that seemed slightly maniacal to me at the time. I watched Dad and Cuan stalk off carefully into the dry bush, crouched over and alert, their rifles close at hand, until they were out of sight. In their camouflage gear, the two of them seemed to be absorbed into the bush.

I was too petrified to move. I was sitting in an open vehicle with two intimidating strangers in the middle of the African bush and my father was nowhere to be seen. I felt utterly vulnerable, although I sure as hell wasn't going to admit it. Could I trust these strange men who barely spoke any English? What else big and hairy, other than them, was lurking nearby?

A gunshot sent a shudder through the bush. Pornai and John immediately leapt out of the dilapidated open four-by-four and beckoned for me to follow. I ran to keep up with them through brittle, knee-high grass. They seemed to know exactly where they were going, but in this maze of twisted acacias and mopane trees I felt completely lost. Suddenly we were in a clearing and there was Dad and a grinning Cuan peering over the body of a dead wildebeest. A trickle of dark blood oozed from a bullet hole in the animal's chest, beside its muscular shoulder.

The trackers ran up to Dad and shook his hand vigorously, patting him and each other on the back. To the trackers, the death of the wildebeest meant that they and their families would have enough meat to eat for a week. For Cuan, he'd bagged a quality trophy animal for his client and it had gone down with one shot, which meant that the animal hadn't suffered. Dad was smiling too, but he wasn't jumping up and down with excitement like the others. He looked profoundly affected by the experience and deeply respectful of the animal on the cracked earth at his feet.

'What do you think, Tammie?' Cuan asked, grinning proudly.

'Oh...it's great,' I lied, forcing a smile.

The truth was, I didn't know whether to laugh or cry. It was the first of many times that Africa would invoke in me this unfamiliar mixture of emotions. Dad and I had argued for many years about his passion for hunting, holding heated debates at the dinner table. Just like him, I'd always been pretty opinionated, but now, in the thick of Africa's reality, I didn't know how I felt about hunting after all.

For the next ten days I joined Dad, Cuan and his trackers on their daily hunts, leaving before dawn and returning in the dark. When an animal had been shot, the men brought it back to camp and hung it up by a massive butcher's hook where it was immediately skinned out in the open. The entire camp staff seemed elated when we returned with a shot animal. I couldn't understand why. The graceful impala, the gutsy warthog, the handsome bushbuck – they were so magnificent in their wild surroundings. Why would anyone want to shoot them?

At the age of fifteen I was an uninhibited greenie and a true animal lover. I used to fundraise for the Worldwide Fund for Nature at our local shopping centre in Townsville, raising money to help save endangered species. At home in sweaty, suburban North Queensland, I'd grown up with an extended family of household pets, from guinea pigs and cats to dogs and rats. Now here I was, in the Eden of large mammals, witnessing animals being hunted by my own father, whom I idolised. And yet the Africans themselves seemed delighted by the killing. It didn't make sense. Everything I'd ever believed in was being turned on its head.

In the west people have become desensitised to scenes of poverty in Africa. A malnourished child with a bloated belly and pleading eyes is a common sight on the evening news, so much so that it doesn't really engender much more than a distant sense of sympathy in many of us. I have found that it is only when you see how little people have in developing countries and how genuinely happy they are with their lot in life that some big questions about your own life hit you.

I felt this for the first time on the day I was introduced to the children of the humble Humani Primary School. On a day's break from the hunting routine, one of the girls working at Humani, Julianne, took me under her wing for a girls-only adventure. I appreciated the change from the routine of blokes, biltong (dried meat) and hunting. I jumped on the back of the motorbike and held on for dear life as she dodged muddy pools and slid over the wet dirt roads around the farm. As we bumped and swerved, Julianne filled me in on the history of Humani.

It all started when Englishman James Whittall, known to all and sundry as Jimmy, pioneered a massive dry tract of no-man's-land in southern Rhodesia in the early 1900s. Of course, it wasn't no-man's-land. It was the home of the Shona people, who were and still are the dominant tribe in Zimbabwe. The Shona people are famous for their smiles within Africa, and for their great sense of humour. The farm Humani was named after a *kopje* or rocky outcrop in the area known to the Shona people as *Gumani*. The new arrivals from England didn't quite get the indigenous name right, but Humani became the Whittall family's legacy. It was land considered useless by the government because there was apparently insufficient water for raising stock.

Jimmy took enormous pride in proving this poor assessment of the land wrong by turning Humani into a prosperous cattle farm. He was buried in the backyard of his old house, which is now the home of Roger, his second son, and Roger's wife, Anne. In the 1970s Roger took an equal amount of pride in proving his father wrong, selling all the cattle and converting the ranch back to game. Slowly he reintroduced animal species such as elephants, black rhino, zebra and nyala onto Humani's 130 000 acres, building up the animal populations to viable numbers. He built a small safari camp called Sambornyai out of rocks, cement, reeds and thatch and offered trophy hunts for plains game to overseas hunters who paid in valuable foreign currency. As the safari business grew more and more successful, Roger built two other camps for both hunters and tourists on the banks of Humani's two wide flowing rivers, the Sabi and Turgwe.

Roger's older brother, Richard, and their sister Jane's husband, Arthur, continued their father's rich farming legacy by growing colourful fields of crops that included mealies (corn), oranges and watermelons. Humani employed hundreds of local Shona people in the agricultural and safari operations and provided their families with homes and food. There was a Humani shop that sold everything from tea and mealie meal to face moisturiser and condoms, and the Whittalls helped build a primary school for the local children. It was a small and thriving metropolis in the middle of nowhere.

To the Shona, Roger attained the African name Shumba, which means lion, a term of immense respect. A stubborn, rebellious middle child with a loud roar that humbled even the most confident of challengers, Roger met his match when he married the kindest of women. Anne was given a Shona name that means the smiling one, and together she and Roger raised a son and three daughters, Guy, who went on to play cricket for Zimbabwe and was something of a celebrity in his home country.

Humani's game-scouts, led by a tall, wizened Shona man named John, patrolled the fences around the perimeter of the farm on a weekly basis, fixing breaks and looking for evidence of poaching. On the occasions when they detected poachers and tracked them down, the culprits were taken to Roger. Humani soon developed a team of game-scouts who made it famous for its impressive wildlife protection, which was probably better than in some of the national parks. Roger's method may have been slightly unconventional, but it worked: he converted poachers to game-scouts. Poachers made the best scouts, he said. To be a good poacher you had to know the bush backwards – the same talent required to become a top game-scout. Roger was turning wildlife killers into wildlife crusaders. Humani's reputation earned it the right to be a custodian for elephants and black rhinos, two species that had been heavily targeted by poachers in the nearby Gonarezhou National Park.

Although I still hadn't worked out how hunting and conservation went together, I knew Humani was a place steeped in history, a place that had seen war and drought, births and deaths, laughter and devastation. It blew my mind to think of what this land had witnessed over the years. I could never have imagined at that point that it would see so much more destruction in the generation to come.

When Julianne and I pulled in at the Humani Primary School, a rain-damaged, painted sign out the front of the single-storey building greeted us with *Humani School Welcomes You. Deeds Not Words Shall Make Us Great.* To one side there was a simple playground with swings and a seesaw made of mopane logs. Beside the playground was a small garden in which the children were growing vegetables.

As we entered the first of the three classrooms, about thirty small, barefoot children jumped to their feet and engulfed me with their exuberant smiles.

'Good mooooorning maaaaadam,' they greeted us in synchronous accord.

The children's eyes followed us around the classroom. The odd nervous giggle erupted as Julianne and I examined the colourful posters on the walls, with words written on them in both Shona and English. In contrast to the shabby clothes of the children, their solemn teacher was dressed in a distinguished grey suit and tie and carried a long, ominous cane as he paced the classroom with undeniable authority. Some of the windows were cracked and the white paint on the walls had long since blended into the tan colour of the dusty outdoors. Torn textbooks that looked as though they'd been printed in the 1960s littered the four wooden desks, each of which appeared to squish ten children onto its long, hard seats. I was overwhelmed and inspired – not so much by the spartan and dishevelled state of the school, but by the uninhibited smiles of the barefoot children. They lit up the room with their excitement. This was a place where little things meant *a lot*.

As we left the building, thanking the teacher for having us, a little boy reached up and touched my arm nervously.

'Excuse me, madam? Please will you take photo of me?' he asked with a grin.

I took his photo and many others and brought them back to Australia with me. By then, of course, I was a different person. In two short weeks Africa had changed my outlook altogether.

To my complete surprise, I had come to understand that not all hunting was bad. Ethical trophy hunting, I realised, was a very different thing from illegal poaching. In fact, trophy hunting, or 'sustainable utilisation', was an integral part of the livelihoods of many people in Africa and it allowed large areas of wildlife habitat to be conserved. I had learned that by targeting the old males of a species that was not threatened, males past their prime and no longer able to contribute to the gene pool, the impact on the population was very low. I had seen how the people of this part of Africa accepted hunting as a normal and necessary part of life. While I went to the supermarket and bought a packaged piece of meat of unknown origin, wrapped up in plastic and polystyrene, Africans hunted in order to eat. Unlike cows squeezed into feedlots and chickens fattened up in batteries, the antelope shot at Humani had at least lived in the wild.

The hunting industry provided jobs for trackers, skinners, cooks, waiters and more, enabling men to provide for their large families. These people *needed* the income and meat that came from Roger's safari operation. They ate every last skerrick of a hunted animal, including the heart, fat and the intestines. There was no such thing as waste. They couldn't afford to. They cooked one meal a day over a campfire, slept on the floor in grass-roofed huts surrounded by goats, and when they had to travel somewhere, they walked, even if it was a journey of fifty kilometres. They had so few material things, but they smiled and joked more freely than any people I'd ever known.

Up until then, I'd not considered my family particularly well off. By Aussie standards we were part of the middle class and Dad worked hard as the director of a property-valuation business to provide for us. I was in Grade Eleven at Saint Patrick's College, a Catholic girls' school, and already I had more education than many of the children at Humani School could ever dream of. Suddenly I was aware of how lucky I was, how wealthy I was in material terms and how content with my lot I should be.

My father said I went to Zimbabwe a child and returned an adult. All I knew was that life for me would never be the same again. My African journey of discovery had begun.

THE DEEP END

Two years later, in 1995, I returned to Humani, not as a safari client this time but as a safari slave. Many long hours of packing shelves and serving customers as a checkout chick at our local supermarket in Townsville had produced the cash I needed to buy the ticket. I'd set my heart on accepting Roger's offer to help his youngest daughter, Sarah, run Humani's safari camps in my gap year between high school and university. Inspired by my first trip to Zimbabwe, I was driven by the promise of excitement and the prospect of gaining some real-life experiences before embarking on a degree. I was almost eighteen.

One thing was for sure: I had no idea where the journey was leading. But the pull of Africa was so strong for me that I had to trust it was leading towards something, even if I didn't know what yet. At the very least it was bound to be an adventure and that was enough for the time being.

Mind you, adventures aren't always fun. Shortly after arriving at Humani I felt awfully homesick and totally out of place. To me and many others, Sarah, my new boss, was terrifying. She was known as Scary rather than Sarah – and for good reason. Tall, thin and with long brown hair, she had a glare that could reduce the most macho professional hunter to a snivelling wreck. She was not impressed with the idea of having to train up a naive Aussie girl fresh out of high school and promptly told me that, at seventeen, I was too young to have a personality. She dragged me along to the camps, ordering me around and generally enforcing her superiority for several weeks before she realised that I wasn't going to give up and go home.

Perhaps I'm a sucker for punishment, but I was determined to make her like me. To be honest, even though Sarah at that time was a self-confessed witch, I couldn't help respecting her. She commanded authority in an environment dominated by men, most of them burly hunters used to women who made the tea, baked cakes and didn't get their hands dirty. No one messed with Scary. I liked that about her. She was tough all right, but out of necessity. Besides which, I didn't believe that she was as hard as she made herself out to be. I'd seen how kind she was towards the family's dogs and cats and I've always felt that you can tell a lot about a person from the way they treat animals.

As we delivered groceries to the safari camps together, made sure that the staff were keeping the camps clean and tidy, and typed letters for Roger on the ancient computer, Sarah developed a grudging acceptance of me. It would grow into much more than that in later years, but I had to survive Sarah's form of initiation first. It was the making of our friendship.

Sarah and I both lived in her parents' house. I never counted the number of bedrooms in Roger and Anne's sprawling, single-storey house, but there were lots, including a verandah enclosed by mosquito gauze which contained several single beds with vibrant-coloured blankets. The house was always full of people, so much so that I think Anne's own children felt like part of an extended family of strangers. Anne loved people and welcomed them into her home like a mother hen.

Her warmth contrasted with Roger's tough authority, but together their energies created an extraordinary atmosphere. There was always a feeling of frenzied activity in their house and it all started at about five in the morning when Roger woke up and began relaying orders or radio calls at full volume. Unlike my own family, in which no one ever raised their voice and I had never even heard my parents argue, in the Whittall house yelling was often the order of the day.

Despite his intimidating exterior, though, Roger seemed to love watching young people thrive at Humani, seeing them grow from naive city slickers into confident adults who loved the bush. From apprentice professional hunters to the occasional foreign volunteer like me, his brand of tough love seemed to bring out the best in people.

I wasn't even close to getting the hang of my duties as Sarah's slave, which were never spelled out particularly clearly anyway, when another job was created for me.

'Miss Matson!' a man's voice called in a deep and eloquent African accent.

His voice was at first lost in the hubbub of the Humani butchery, where I was helping Sarah organise some meat for clients staying at Sabi Camp. We were filling a cool box with impala roasts, eland steaks and warthog chops. Engorged flies buzzed around the chilly room, the off-white walls splattered with blood and entrails. Skinned legs and rib cages of shot antelopes hung from hooks in rows.

'Miss Matson!' the man called again politely but louder this time.

I turned away from the hanging carcasses to see a Shona man outside the screen door with a smile much too large for his thin face. Despite the steamy weather, he was dressed in a suit and tie.

He stretched out his hand to me and said engagingly, 'Miss Matson?'

Suddenly it hit me who he was and I smiled. He was the principal of Humani Primary School, with whom I'd been corresponding from Australia for the past year.

'Mr Hunde?' I registered.

'Yes, yes! It is me! Oh Miss Matson...' he exclaimed exuberantly, shaking my hand violently.

'Tammie is fine, Mr Hunde. My friends just call me Tammie.'

He looked a little shocked at this familiarity but continued animatedly, 'Tammie... I am so happy to see you. When one of the teachers told me you were here I came back early from vacation to see you. How long are you staying for?'

'I'll be here for the next six months,' I answered, still getting used to the idea. I had never been away from home for more than three weeks at a time, and never as far away as Africa by myself.

'Six months? Oh, that is very good, very good indeed! But not long enough for us, I am sure.'

Mr Hunde launched into what seemed to be a half-prepared speech, thanking me profusely for the stationery I had sent him for Humani School from the fundraising efforts of my school, Saint Patrick's.

He paused for a breath, which gave me the chance to ask him how he was. He looked at the ground, suddenly quite miserable, and explained that he was not happy at all. Accommodation at the school was poor, transport was even poorer, Humani was a long way from his family in Zaka and, on top of that, there was no beer.

I empathised with him and, to change the subject, suggested that I could come and help with teaching at the school if he wanted. I explained that I wasn't qualified but that I'd be happy to do it in my spare time.

'Miss Mat– Tammie,' he corrected himself. 'We would be *very* grateful, most grateful indeed!'

A few days later I visited the school with Anne to drop off a load of textbooks and posters from Saint Pat's.

'It gives me so much pleasure to have you here, Tammie,' Mr Hunde declared as he accepted the boxes. 'Now, would you like to come and meet your class?'

I just about fell over in fright. I'd offered to help out, but I certainly didn't feel capable of taking a class of my own. I'd only recently finished school myself! But with great enthusiasm Mr Hunde was leading me to one of the classrooms. I wanted to object. I had no idea how to be a teacher. My hands started sweating and I felt my face burning up in panic.

I protested desperately, 'Mr Hunde, I'm not a qualified teacher! You know that, don't you?'

'Ja, it is not a problem,' he replied, brushing my doubts aside. 'This is Grade Six and Seven. Their English is quite good.'

And with that he led me into the dark classroom and into my sudden and unexpected career as a primary school teacher.

'Good moooooorrrrnnning maaaaadam,' a dozen barefoot Shona children chanted in unison.

I took in my humble surroundings. The school had no electricity, so there were no lights in the classroom. A few more windows had been smashed since I was last here, and a couple of the long wooden seats were broken. A blackboard on the front wall displayed the day's work in white chalk. A goat poked its head around the door, only to be shooed out by the little girl nearest the doorway. It was pretty much as I remembered it. The children were smiling with the same healthy, glowing faces that had inspired me last time, as if they'd swallowed light bulbs, illuminating the bleak classroom. Suddenly, watching their shy smiles and expectant eyes, I realised that these kids were as anxious as I was. We were all as freaked out as each other. That made me feel a little better.

'Class, I introduce you to Tammie Matson,' Mr Hunde boomed. 'She is the kind lady who sent us letters from Saint Patrick's.'

And with that, to my horror, he handed it over to me. I stood there for a moment, not knowing what to say.

'Um…hi kids,' I ventured, 'Have you been working hard today?'

Fifteen shiny black faces were all eyes and no mouths. They looked even more petrified than me.

I smiled and tried again, speaking more slowly this time. An uncomfortable silence lingered in the room. I wished it would swallow me up and spit me out into another universe.

Mr Hunde looked very embarrassed. He asked the question again. Immediately, his words commanded a response. A little boy at the back of the room shot his hand up in the air and yelled something I couldn't understand that started with 'Madam'.

'Why do you not stand?' Mr Hunde roared.

The boy leapt to his feet as if zapped on the bottom with an electric prod and called out his answer again. But I still couldn't understand his hurried speech, so I turned to Mr Hunde for help.

'He says he learns arithmetic,' Mr Hunde explained graciously.

I turned back to the boy and asked him another question that was met by another stunned silence.

Mr Hunde interceded. 'It is just that they are not familiar with you.'

I wondered how I would ever be able to teach a class that couldn't understand me, nor I them. On top of that, they appeared to be terrified of me! I had expected shyness, but not fear.

The Whittalls agreed to let me teach at the school in my spare time. They seemed baffled as to why I would want to, but as it didn't interfere with my work in the safari operation, they supported it. I realised that the invisible lines between people of different skin colours in Zimbabwe meant that I was probably the first white person ever to teach at the school.

Mr Hunde suggested that, for starters, I teach English for a couple of hours each afternoon after school had finished for the day. I agreed despite my self-doubts and tried to overcome the butterflies in my stomach.

On my first official day of teaching I asked Mr Hunde if I could watch him for the first hour so I could get a feel for the way the classes worked. After an hour of sitting in the back of the room and listening to Mr Hunde's lesson, I still felt as raw as uncooked meat. And I was about to leap into the mincer.

'Are you ready now?' Mr Hunde asked me, his eyes twinkling.

With fifteen expectant faces awaiting my response, I knew I couldn't put it off any longer. I forced all my uncertainty aside, squeezed out from under the desk and walked up to the front of the room. Flicking through an ancient textbook, I found a vaguely interesting article on the San Bushmen of the Kalahari Desert. The rest of the book was filled with stories about little English children, a product of British colonial rule when Zimbabwe was still Rhodesia. I wondered how these children could possibly relate to such foreign content.

'Okay!' I smiled, trying to sound like I did this sort of thing every day of the week. 'Who can tell me something about the Bushmen?'

I took care to speak very slowly because by now I suspected that Mr Hunde's estimate of the children's English comprehension was slightly optimistic, not to mention that they had probably never heard an Australian accent in their lives. Nonetheless, the question met with a deathly silence.

'Does anyone know about the Bushmen?' I asked again.

A couple of the girls giggled shyly. Mr Hunde sat in the corner with that embarrassed look on his face again. It didn't appear that I was going to get any answers, so I kept talking. The more I talked, the more relaxed I became, until it didn't seem to be my voice doing the talking at all.

'In Australia, we have the Aborigines. The Aborigines were the first people to live in Australia before the Europeans came. In Africa, the first people here were the Bushmen. This is a story about the Bushmen who live in the Kalahari Desert. Who would like to read?'

I paused and willed for someone to answer. To my amazement, a hesitant hand rose.

'Yes, what is your name?' I smiled thankfully.

'My name is Benjamin, madam,' the boy said shyly – in perfect English.

'Thank you, Benjamin. Would you like to read?'

By the end of the first page, half of the class was looking out the window and even Mr Hunde had become bored and left the room. These kids were normal after all!

I asked the children to put their books down. 'Okay, class. I am from a place called Australia. Australia is far, far away across the ocean. I am here to help you learn English. My English is good, but my Shona is poor. I want to teach you English, but I also want you to teach me Shona. I think we can both learn. Can you do that?'

At last the room broke out in giggles and grins all around. The first barrier had been broken. I asked Benjamin to carry on reading for a little while longer and when he finished I said, '*Ndatenda* Benjamin.'

Ndatenda – thank you – was the first and most useful word I'd learned.

One of the boys ribbed his mate with his elbow and I heard him whisper in astonishment, '*Ndatenda!*' as if I'd just sworn in the presence of the Queen Mother.

I left the school that day feeling as though I was embarking on a thousand mile swim through shark-infested waters, but at the same time filled with excitement. I'd faced the sharks and lived. I could do it again. I wasn't a teacher, and both the kids and I knew it, but that didn't mean we couldn't make a difference in each other's lives.

Word travelled fast through the Shona compound that there was a new teacher at the school and undoubtedly the fact that I was a 'whitie' was fuel for village gossip. To get to school each day I would walk on a dirt track that weaved through dozens of grass-roofed huts, past smoking campfires and herds of bleating, patchwork goats. Within days, it seemed that half the Shona population of Humani knew who I was. Suddenly I was a celebrity.

'Taaaaaamie! Allo Taaaaamie!' little children would call to me as I walked by their huts, waving and running out to greet me, often skipping along behind me and offering to carry my schoolbooks.

Some of them carried their little brothers or sisters on their hips, although they couldn't have been more than five or six themselves. One of the boys in my class, a serious but kind fellow called Obert, waited for me to pass his family's hut before rushing out, shooing the little children away with authority and taking my bag as if it were a great treasure and an honour.

When children didn't escort me to school, Rosalind, a tame eland that Anne had raised from an infant, followed me. An eland is a large tawny-coloured antelope with horns that resemble twisted liquorice sticks. Physically Rosalind was an eland, but mentally she considered herself a goat. She spent her days grazing with her goat family around the huts. Spotting me, she'd often wander over and give my fingers a hello suck.

On one day, while I was talking to Rosalind on my way to class, a man called out to me from a nearby campfire. '*Ticha! Eh ticha!*' He threw one leg over an invisible horse in imitation of a jockey and pretended to ride. He was trying to tell me that it would be much quicker to ride Rosalind to school than to walk each day. I thought about it for a second, then took one look at Rosalind's horns before putting the idea out of my mind. I was making enough of a scene as a white teacher without riding a bucking eland to school as well.

On my third day I was technically the most experienced teacher at the school. I was beginning to understand why Mr Hunde had been so keen to have me there. He had been admitted to Chiredzi State Hospital, an hour's drive away, with a bout of malaria. Muvhu and Mr Sithole, two other teachers, were nowhere to be found and Junior, who taught Grades One to Three, was allegedly in Mkwasine gaol, having been arrested for thumping somebody. The only teacher at the school other than me was a newcomer called Makai who had been sent in to replace Junior until he got out of gaol. I also discovered that, having completed Year Twelve, I was actually better qualified than most of the teachers at Humani. None of them had studied beyond the equivalent of Year Eleven. This gave me a much-needed boost of confidence.

A group of boys often stayed after class while I marked their spelling. One day I asked them how old they were. Clever announced that he was twenty-five, Barire seventeen and Obert fifteen. I thought they must be joking, but when I asked Muvhu about it later, he told me that they were probably telling the truth. Clever, I discovered, was the oldest at a ripe twenty-three, although like all of the 'kids', he looked to be in his mid-teens;

and the average age of my class was fifteen. This was due to what Muvhu described as 'family problems'. Perhaps a parent had died and school had had to take second place to household chores and minding younger siblings.

Then there was Benjamin, the boy who had shown the courage to answer my question on my first day of teaching. He was an orphan and lived at the teachers' quarters under the watchful eye of his guardian Mr Sithole. Although Mr Sithole was a teacher, Benjamin only attended my classes about twice a week. Often I would see him cleaning Mr Sithole's motorbike during my class. There was something about Mr Sithole that I didn't trust. The sleazy, too-nice way he spoke to me reminded me of a politician.

As I grew to know the kids, I began to realise that a couple of them had serious learning problems. One little girl, Athania, always appeared to listen carefully in class, but I'd caught her copying one of the other girls' work a few times. Her written English made no sense whatsoever.

'What does this mean, Athania?' I asked her one day, pointing to a sentence she had just written.

She didn't say a word, only looked up at me with apologetic, puppy-dog eyes and a faint, almost scared smile. I realised then that Athania barely understood a word of spoken English and she certainly couldn't write it.

Like Athania, Paul also produced work that I couldn't read. He would string letters of the alphabet together randomly in a line so that they looked like sentences but in fact made no sense whatsoever.

At first I thought he might have been writing in Shona, but when I asked Muvhu about it he said matter-of-factly, 'No, this is not Shona. He is writing rubbish.'

'So how did he get to Grade Six when he writes like he should be in Grade Two?' I asked, incredulous.

Muvhu replied with a resigned shrug, 'Social background is a major factor...and his hearing is not good.' He gestured to his own ear with a pained expression.

Hearing loss was a valid explanation for a lack of literacy, but social background?

Muvhu seemed hesitant to explain. It took me a long time and a lot of prodding to coax an answer out of him. Both Paul's and Athania's parents were *nyangas* or witchdoctors. Because they practiced *mshungu*, black magic, no teacher in the past had been game enough to keep their child down a grade.

'The parents... they are... ahh... difficult,' Muvhu said carefully, as if afraid that a *nyanga* might be listening. He adjusted his glasses nervously. It seemed that even Muvhu, a well-educated teacher from the capital Harare, was unwilling to cross a witchdoctor.

Later, when he'd returned from hospital, I asked Mr Hunde about *mshungu*.

'Tammie,' he said solemnly, 'you once asked me why I don't like it here at Humani. That is why.'

Mr Hunde had not been born at Humani and was considered an outsider, so he had been the target of *mshungu* many times.

'Let me explain,' he continued. 'I am walking to the butchery. I will walk with you.'

As we walked along the dusty track coiling its way through the village, Mr Hunde tried to explain how *mshungu* worked.

'That is the home of the *nyanga*,' he said in a quiet, reverent voice as we walked past three particularly well-maintained thatched huts. Each was extravagantly painted in triangular patterns of maroon, black and mustard. The bare earth around them had been immaculately swept with grass brooms.

I tried without much success to pronounce the word *nyanga* as Mr Hunde did. There was a kind of click in it when he said it. He picked up a stick and wrote the word in the dirt with it.

'Tammie, this is why I am always sick in this place. It is the *mshungu*.'

Mr Hunde had been appointed as the Humani headmaster by the regional school authority, but prior to his arrival Mr Sithole had been the principal. Mr Sithole had lived at Humani his whole life and now he'd been demoted to a teacher role. Unfortunately for Mr Hunde, Mr Sithole just happened to be a low-ranking *nyanga*. Mr Hunde believed that the reason he was constantly coming down with bouts of malaria was because Mr Sithole was always cursing him.

All it took was a cursed item, like a leaf or a strand of hair, to be laid on the path along which you, a devout believer, walked and that was it. Your fate was sealed and you believed it absolutely, even unto your death, which usually came swiftly thereafter. The magic could kill a toddler or make a man gravely ill or cause a woman to lose her unborn baby. Black magic was a matter not taken lightly by anyone and *nyangas* were taken very seriously indeed.

Roger had always detected thieves by consulting one of the local witchdoctors. He would simply line up all the staff for the *nyanga* to examine and even before he'd had a chance to work his *mshungu* the culprit would inevitably stand up and hand himself in. Rather face the retribution of Shumba than that of the *nyanga*.

Some white people proclaimed that black magic was all a load of superstitious nonsense but everyone at Humani knew that with the right mixture of herbs, maybe a vervet monkey's teste or two, and a strand of human hair, the *nyanga* could induce the death of anyone he wanted to as quickly as he liked. Of course, there was never any evidence of his influence, because most of the time the victim would die slowly, often alone, resigned to his own fate. Some would say it was malaria. Others that it was AIDS or TB. But everyone knew. It was *mshungu*.

I was aware of the *nyangas*' status in the community, but the theory was that we whities weren't affected by their magic because our bearded old God on his puffy cloud told us we weren't allowed to believe in it. That didn't stop the *nyanga* finding other ways to exert his influence, but never could I have imagined that a traditional dance, performed by the schoolchildren for some American clients, would induce the *nyanga* to act with such malice.

I had approached Mr Hunde with the idea a few days after our *mshungu* conversation. I explained to him that it was very likely the Americans would leave the school a tip which could be used to buy Shona–English dictionaries. Mr Hunde was thrilled by the prospect and began organising it immediately.

After weeks of preparations, the dance was held just before sunset in the school grounds. The clients sat on comfortable canvas chairs under the shade of a tree, the sun setting behind them. Several older boys and adults stood or squatted in a semicircle, sacred drums made of carved wood and hide perched in front of them. A lot of villagers had congregated to watch and the anticipation was palpable.

The pounding of the drums was started by one of the teenagers, a boy clad in an old T-shirt and torn shorts. Tapping lightly, then more heavily, then more heavily still, he filled the bush with the beat of his single drum. Then abruptly he stopped. A rivulet of sweat trickled down his forehead and splashed onto the skin of his drum. In the silence he looked up at the older man beside him and grinned. Suddenly everyone started pounding together in a rhythm that seemed to emanate from the very heart and soul of Africa. Each drummer pounded his drum in a style unique to himself but contributed a perfect complement to the great swell of the rhythm.

A middle-aged woman from the village threw her hands into the air and began wiggling her large buttocks, her great breasts heaving under her floral dress. Her face pointed skyward, her eyes shut in ecstasy, she began to ululate, a haunting, high-pitched wail which sent a shiver through the crowd and prompted a wave of ululations among the women.

When the children emerged from one of the classrooms, the ululations increased and everyone began clapping. Wide, toothy grins lit up the children's faces. The girls came on ahead, their eyelashes fluttering, both seductive and shy. They gyrated their hips, their arms undulating like serpents above their heads, their buttocks wobbling in time with the drumming. Then came the boys, each carrying a long wooden stick as if to imitate spears. Their faces were those of young warriors, stamping the earth, jumping boisterously before hoisting their sticks aggressively and growling at the audience. Their ankles were ringed with black and white spotted guineafowl feathers. With every stamp of the boys' feet the feathers whipped up as if in flight and tiny puffs of dust exploded from the earth.

The pounding continued for an hour, by which time the drummers had plunged into a trancelike state, oblivious to the sweat pouring from their skin. After the children had performed their routine, everyone in the audience leapt up and joined in the dancing. Even one of the client's wives stood up in her Dr Livingstone safari wear and began flicking her bleached blonde hair around and stamping her feet with great relish. Something in the drumming caught in one's stomach, pirouetted into the chest, then lurched its way to the extremities, with the result that it took a fight against nature to stop yourself moving to the rhythm. It wasn't that you simply felt the music. You *were* the music. The primal urge was contagious. Nobody wanted it to end.

Afterwards, the clients made a donation to the school in thanks. The amount was enough to purchase about twenty new dictionaries. In front of the excited crowd, the money was given to me and I announced that it would be used to help the children learn to read and write. A round of excited applause and ululations rang out from the villagers.

I noted the old man Hama looking a little solemn, but I thought nothing of it. He was the father of Athania and he'd led the drumming. Earlier in the day, Anne had ordered some of the men to take a kudu carcass from the butchery and to give it to the participants in the dance. The village would feast tonight. The clients were escorted back to their camp, although the drums continued pounding in the village until the early hours of the morning. The evening had been a huge success.

Or so I thought. As it turned out, Hama had other ideas for the Americans' donation, although I did not discover this until many days later, when Mr Hunde appeared at the gate of Roger's house. He looked wretched. His face was slick with sweat, and his eyes were yellow and bloodshot – a clear sign of malaria.

'What is it, Mr Hunde? Are you sick?' I asked him.

He shook his head sorrowfully. 'I am sick, ja. Tammie, I have a problem. Please, I ask you to help me.'

'What is it?'

'The money. The money from the Americans. You know. That money must please be going to Hama. He says that if you don't give him the money, I will die.'

'What do you mean, *die*? You're probably getting malaria, Mr Hunde. Let me give you some *muti* (medicine) and you will be fine.'

He shook his head vehemently, fear in his eyes. 'No! Please, Tammie, you do not understand! It is the *mshungu*. If you do not give the money to Hama, I will be dying. There will be no more school. Please, you *must* help me!'

It was evident that Hama was using my friendship with Mr Hunde to get to the money. It was blackmail but I had no doubt that Mr Hunde would indeed die if I didn't intervene. It was going to take a little more than quinine to heal this type of malaria.

I explained the problem to Roger, who didn't even blink an eyelid.

'Anne, call for Hama,' he commanded gruffly. 'Tell him I want to see him here.'

Hama arrived at the gate several hours later, escorted by a couple of voluptuous wives wobbling subserviently along behind him. A rusted imitation Rolex watch hung loosely from his skeletal wrist, not much use for telling the time but a highly fashionable status symbol.

'Hama, what is this all about?' Roger began.

The two men stood face to face, only a metre apart. Eye to eye, Hama wasn't much shorter than Roger. The old man's skin hung off him like clothes from a washing line, but that didn't make him any less formidable. It was a battle of wits and both men knew it. I couldn't understand their words, which were spat and cursed in Jalapalapa, the hybrid language of Zimbabwe that is a mixture of Zulu, Shona, English and a few other languages. Back and forth the men argued, louder and louder. The old man's voice croaked aggressively and he appeared to be hurling abuse at Roger, but Roger was equally stubborn and authoritative.

Later, Anne translated it for me. 'Hama told Roger that he must not interfere with these things and that the people don't need the money for the school because there are other things to buy.

'Roger said, "Hama, this money is for the children. Do you not have children at the school? Do you not want your children to learn English so that they can get jobs and make money for your family?"'

I recalled that Hama had been shaking his head vehemently, his eyes flaring up in anger.

'Roger said to him, "You *will* remove the curse, Hama, or you will leave Humani at once, you and all your family. There is no question."'

I'd watched as the old man took a step back, a look of shock on his face, as if he'd just been punched. For a second, he seemed to be looking through Roger, as if seeing something that human eyes cannot. There was a long pause. All of a sudden he smiled, a great yellow, gappy grin. He chuckled, then coughed and spat on the dirt. And as fast as it had started, it ended. Mr Hunde would live to see another day – until the next *nyanga* took a disliking to him, that is.

I began to see improvements in the children's work at school. As my Shona improved, so did their command of English. Nonetheless, sometimes we had to resort to sign language. During one lesson one of the boys, Barire, stood up and said desperately, 'Excuse me, Tammie?' He held his nose with one hand and with the other he brushed the air in front of his nose as if to dissipate a foul smell. The class was in hysterics.

'You want to go to the toilet, Barire?' I asked, smiling.

'Yes! Thank you!' he exclaimed, sprinting for the door.

Sometimes, no matter how hard I tried, I couldn't get the message through.

'Excuse me, sir?' said Daidzirai.

'Daidzirai, my name is Tammie. For a man, *murume*, you say "sir". For a woman, *mukadze*, you say "madam". So you can call me madam or Tammie, okay?'

'Okay,' he replied, as if he knew exactly what I was talking about.

Five minutes later, he put his hand up and said, 'Excuse me, Sir Tammie?'

Our classes were never without amusement. In one lesson a little girl of about seven appeared at the classroom door. Suddenly Clever stood up, looking sheepish. Without invitation, the little girl yelled at him in Shona with a voice far too large for her size. It sounded like an order. The class broke into wild laughter. Clever, clearly embarrassed, brushed her away as if she was a pestering insect. Undeterred, again she yelled at him in a shrill, bossy voice and then stomped off.

Later I asked Muvhu what the little girl had wanted.

'She is his sister,' he explained, chuckling. 'She was telling him he must go to Chibuwe to buy some beer for their father. Clever told her, "Go away! I am at school!" Then she said that he was going to be in big trouble when she told their father.'

One of the brightest boys in my class, Tadius, had to leave half an hour earlier than the other children because he lived about ten kilometres from school. This wasn't considered an unusual distance for a child to walk each day to receive an education. Walking home late one afternoon he'd walked into a herd of elephants but had managed to get away with his life. These children considered school to be such a privilege, even wild elephants couldn't keep them away.

Gradually I discovered that all the children had distinct talents. Those who didn't excel at English writing could draw beautifully or were great at sport. I developed a soft spot for each one of them. Athania, Hama's daughter, really couldn't give a damn about schoolwork, but she simply loved being there. She wiped down the blackboard with a duster each day as if it was a priceless work of art, swinging her hips and singing sweetly as she dusted. Daidzirai and Tadius, the two boys who competed for top-student status, showed as much rapture upon receiving a silver star sticker for getting a hundred percent in spelling as athletes would receiving gold medals at the Olympics. Taadini would draw skilful sketches of animals in the time it took the others to pick up a pencil, and kind-hearted Obert never failed to carry my books to school each day.

Mr Hunde surprised me one afternoon when, out of the blue, he became quite emotional and said, 'Tammie, the people are asking about you. They want to know what you are doing at the school and why you are doing it. I can only tell them this. It is love... You have this *love* for the people.'

I was shocked to hear this. After all, I was the main beneficiary of my time at the school. My pupils were great teachers – I was learning as much from them as they were from me. Besides, we were virtually the same age as each other!

After two months of teaching and working at Humani, my life took another unexpected turn. Roger needed some help at his safari camp in Zambia and, not knowing what I was letting myself in for, I offered my services. I was about to throw myself into the deep end again, but my time at Humani School had taught me something important. You only truly know how much you are capable of if you have a go at things you don't think you can do. Right now I felt ready for anything.

THIS IS ZAMBIA

Like many African countries, the name 'Zambia' has a nice ring to it, sounding as though each syllable is being beaten from a drum. Zam! Bi! Ah! After two months in Zimbabwe I thought I was ready to take on anything Africa had to show me. But to be honest, I didn't know the first thing about Zambia.

'Do you need a lady's touch at your camp in Zambia?' I asked Roger at lunch one day.

He didn't look up from his meal of cold impala roast, green salad and fresh farm mealies, golden and steaming. 'Well, we always need people to work in Zambia,' he replied, a note of disbelief in his businesslike tone.

Peter, one of Roger's apprentice professional hunters who was about my age, had simply laughed when I'd first suggested going up there. Mind you, he and another Humani bachelor, Mark, had once told me that women were only given legs to walk from the kitchen to the bedroom, so I took Peter's response with a pinch of salt. If anything, it spurred me on to know more about Zimbabwe's poorer northern neighbour.

'In our brochures we call Lionheart Camp "rustic",' Sarah said. 'Use your imagination, Tam.'

'If you need someone to help out up there, Roger, I'm happy to go,' I suggested without batting an eyelid.

'Zambia?' Roger responded, sounding more incredulous by the second. 'It's really rough up there, Tam... *Really* rough.'

He smiled. It never ceased to amaze me how genuine and soft a smile could emerge from this gruff, hard man. It was like watching a lion turning into a teddy bear.

'Yeah I know, but I'm an Aussie, remember? Aussies are known for doing stupid things.'

'You *want* to go?' He seemed puzzled that a woman would even consider the possibility, let alone ask for it. He was probably wondering which one of the guys up there I had a fancy for.

'If there's a job,' I replied, 'then, yes.'

Roger paused in contemplation. 'Actually,' he announced suddenly, 'it'd do you good to rough it for a while... Get bitten by a tsetse or two... Might toughen you up.'

Several weeks later I was informed that I would be driving up to Zambia with a hunter by the name of Hilton Nichols. Hilton was a celebrity in the Zimbabwean hunting circuit and clients were booked to hunt with him for two years in advance. I had an image of a tall, broad-shouldered, rugged-looking fellow with the eyes of a leopard.

So I was surprised when confronted by a stocky bloke of average height with a happily rotund belly, a neatly clipped head of straight brown hair, and old-fashioned black-framed glasses. The amusement in his deep brown eyes far from mirrored the first gruff words he said to me.

'Are you ready to go?' he grunted. 'I want to be gone in five minutes.'

So began our two-day long, dusty, bumpy journey to Zambia. As always, half the adventure in Zambia would be in getting there.

Hilton's single-cab Land Cruiser was heavily loaded with supplies for the camp. Crates of booze, boxes of tins, and bedding jiggled and shook as we lurched over bumps and narrowly missed gaping potholes in the main bitumen highway. Hilton chain-smoked strong cigarettes as he drove and chatted good-naturedly. He was a far cry from the great white hunter I'd imagined, in fact he was a total softie.

'Now this is a really *muntee* city,' Hilton murmured with a sly slant to his smile as he pulled the Cruiser off the highway and into the township of Karoi in northern Zimbabwe. This was the home of Hilton's tracker, Mudzingwa, who was coming with us to Zambia.

A wide dirt road threaded through the crowded town, lined on both sides by hundreds of conical thatched huts. The smoke of innumerable small campfires rose in white tendrils to merge with dust-thick air. Hilton pointed out the homes of the more affluent residents: red-brick, two-roomed houses with tin roofs, packed closely together with less than a metre between. Hordes of people ambled casually into the middle of the dusty road as we passed through. There were people *everywhere*. People walking, people talking, people picking their noses and scratching their arses, a vast wave of people in a heaving sea of dust.

Hilton turned sharply onto a skinny dirt track and pulled up at a square red-brick dwelling. A neglected bed of violets lined the earthen path to the door. A small barefoot girl danced out excitedly when our car stopped out the front. She and Hilton conferred briefly in Jalapalapa. I noticed several inquisitive faces poking out from behind the neighbours' doors and through the holes in the dilapidated brick walls that acted as windows. I gathered it wasn't every day that two whities in a Land Cruiser popped in for a visit.

'Should've known…' Hilton commented wryly as the Cruiser growled back out onto the main road. 'If there's a pub down the road, that's where he'll be.'

We pulled up beside a tall, closed-in fence with the magic word *Chibuku* – beer – painted on it. Rowdy, drunken male voices signified that this was the action centre of the township: the local pub. Rural Zimbabweans usually brew their own *chibuku*. It's cheaper than buying the bottled stuff. This potent home brew resembles muddy creek water but packs a more solid punch than regular bottled beer because it continues to ferment once in the stomach. As a result, you keep getting drunker and drunker long after you've stopped drinking.

'I better go and find him.' Hilton's tone was serious. 'You should get out and keep a *very* sharp eye on the vehicle.'

The back of the open vehicle was loaded up with more groceries than most Zimbabweans would buy in a year. I got out of the car tentatively and, perching one hand on my hip and leaning the other on the back of the Cruiser, pursed my lips into the meanest, toughest expression I could manage. Several men with the glazed expressions of too much *chibuku* drifted past me, making unsubtle examinations of the Cruiser's contents and its short-arse defender. They skulked around the vehicle like hyaenas circling a kill. I tried to stand taller and look meaner.

'Hello madam,' one slurred, approaching the car.

He wasn't looking at me directly as he spoke but was eyeing the contents of the vehicle. I glared at him and clenched my jaw tightly. I was trying desperately to appear fierce and unafraid, but the men were stone drunk and I felt very small and vulnerable.

'How are you?' the man continued.

He was moving in closer, moving in for the kill. I stared him right in the eyes in a feeble attempt to be intimidating, but he came closer and closer until he was standing right beside me.

'Where is your husband?' he purred.

The others were circling on the other side of the vehicle, looking for an opportunity to grab something that wasn't tied down. I was in trouble. I couldn't watch him and the others at the same time and I knew they were setting me up. I felt like a mother impala fending off a wild dog while the rest of the pack devours her youngster. Hurry, Hilton!

'He has gone inside. He will be here in two minutes, *now now*,' I spluttered, then quickly turned to check that one of his dodgy mates wasn't swiping something from the car while this chap was diverting my attention.

'He is coming now,' I said again, sounding more confident than I felt.

Suddenly the man spun around and stumbled away, his mates staggering behind him. At first, to my delight, I thought that my hard-core demeanour had put them off. Yeah! But then I turned to see Hilton and Mudzingwa emerging from the pub, scowling threateningly at the gang of drunks as they retreated like puppies with their tails between their legs. My pride was wounded momentarily, but secretly I was immensely grateful for Hilton's impeccable timing.

My first impression of Zambians was that they had a lot of faith in the morality of highway drivers. Villagers holding out the hairy, pale green fruits of baobab trees, unusual rocks and anything else that a tourist might buy stood smack bang in the middle of the highway in a strange game of chicken. On several occasions, while sounding the horn and slamming on brakes, Hilton swerved at 120 km/h to avoid driving right over them. Teenage boys in ragged shorts stood beside the road and made shrill, piercing whistles as we drove past, and one even attempted to spit in the passenger window. Dusty villages dotted the dry, barren landscape, blending perfectly into the brown earth. We dodged foot-deep potholes the length and breadth of the Cruiser on what was Zambia's main highway. Zambian roads, I discovered, varied from very bumpy to so bumpy that you were better off driving on the dirt edges.

By the time we arrived in the capital city, Lusaka, it was late at night. Mudzingwa was sleeping on a swag in the back of the Cruiser (possibly sleeping off a hangover). From the car the sprawling city seemed smelly, dark and dingy. I felt as though I needed to be on my guard, but that wasn't much different to most big cities that I'd been to in Africa. Admittedly, we were exhausted and hungry so I may not have seen Lusaka in a fair light.

'We've got a problem,' Hilton announced ominously.

'The only problem I've got is in my belly,' I replied.

My stomach growled in response.

'Me too,' Hilton replied. 'That's the problem. We haven't got any Zambian kwatcha and none of these stores take Zim dollars.'

'What about US?'

Hilton shook his head in frustration. The banks were closed and although we had several currencies and bankcards between us, none of these meant anything in Zambia.

Suddenly Hilton's face lit up. A vehicle had pulled up at the dimly lit service station where we were parked and a white guy, the first one we'd seen in Zambia, slid out of the driver's seat and called for service. Without wasting a second, Hilton introduced us both to the man and explained our sorry predicament.

The stranger shook his head. 'Sorry, man.'

Hilton drooped.

'I wish I could help you but I'm on my way out of the country and I've hardly got a kwatcha on me,' the man explained.

Just then two Zambians dressed in smart business suits and ties joined the conference.

'Do you want kwatcha?' one of them asked Hilton.

Hilton's surprise quickly turned to scepticism. I could hardly blame him – I'd seen the two stylishly dressed men get out of their Peugeot ten minutes earlier and there had been something about their suave appearance that had reminded me of the mafia.

'How much? Fifty? A hundred?' the men offered.

The white man interjected with suspicion. 'There's a motel down the road about a kilometre back towards the city. Maybe you want to try there.'

Hilton nodded.

'Ja! We will show you the way!' the Zambians offered eagerly. 'Follow us. It is not far.'

I couldn't help but wonder why the two men were being so friendly. Obviously they wanted something out of this, but what? We followed their dilapidated, push-start Peugeot to the motel. Hilton thanked them, hinting that we would be okay now and no longer needed their help, but, undeterred, they followed us inside.

'Do you have any kwatcha?' one of them asked the motel attendant before Hilton could say a word.

The attendant thought about it for a moment, glanced at the two men, and then, as if making a decision based on what he'd seen in their faces, shook his head, saying they had run out. That a motel in a capital city would run out of the national currency seemed unbelievable. But of course this was Zambia, where I would learn that a westerner's idea of unbelievable is equivalent to a Zambian's idea of normal.

'We will help you,' our suited heroes came to our rescue.

They gestured that we should follow them and, to my abject horror, we ended up in a dark parking area with no one else in sight.

'How much do you want?' one asked.

The men began the exchange and soon Hilton had in his hand a three-inch thick wad of Zambian kwatcha.

We pulled away in the direction of the service station where earlier we'd been enticed by advertisements for soggy chicken burgers and oily chips.

'I think we've just been seriously had,' Hilton commented with a hapless smile.

But to our amazement, the kwatcha we'd been sold appeared to be valid. Soon our bellies were full and we were feeling a lot better about Zambia and life in general. What was even more amazing, though, was that our opportunist rescuers returned shortly after in their Peugeot because we had actually accidentally short-changed them!

After an uncomfortable night spent snoring in the front of the cab, our exhausted, bruised and battered travelling trio rattled into Malambwe Camp. It suddenly appeared out of nowhere from behind a screen of thick bush that looked exactly the same as all the rest of the bush we'd been driving through for hours. A small aggregation of reed and thatch huts and canvas tents simply materialised out of thin air. Civilisation.

Roger had described the camp as 'seriously rough', but at this point, after days of driving, nothing could have been more alluring. Set on a small tributary of the Malambwe River, the camp was surrounded by a two metre tall wall made of dried grass, which was held up by wooden stakes and wire. It was a bush camp, complete with a long-drop toilet that consisted of a plastic seat perched on top of a red-painted barrel. It certainly wasn't luxurious, but it was authentic. A combination of small grass and reed huts for the staff and a large canvas tent with an ensuite bathroom for the clients made up the basic accommodation.

A barefoot, unshaven apparition in a shabby T-shirt appeared before us with greetings. It was Mushie, Hilton's nephew, the manager of Roger's two camps in Zambia. He was surrounded by a variety of flies. Dozens of tiny black mopane flies wasted no time zeroing in on the moisture sources of the new bodies in camp. One flew into my eye and drowned in lachrymal fluid before being hastily wiped out. A handful of the bulging winged soldiers known as tsetse flies, wielding blood-sucking needles that carried sleeping sickness, a tropical disease that causes extreme lethargy and even death, had been on the attack in the cab of the Cruiser ever since we'd entered 'tsetse country'.

I'd been shivering in the car the night before, but by midmorning the winter day had warmed up to the low thirties. I was sweating, exhausted, being eaten by tsetses, harassed by mopane flies and, after very little sleep, felt like hell in a heatwave. I had some sense now of what Roger had meant by 'rough'.

Mushie ushered me into one of two staff huts made of reeds and grass.

'You'll have to bunk in with Peter and Mark,' Mushie said with an amused smirk.

I laughed, thinking he was joking. No worries Mushie, I thought, I'll just act like I sleep in the same hut as two female-deprived, eighteen-year-old, hormone-pumping males in the middle of Woop Woop every day of the week. I was fresh out of an all-girls Catholic school. Mushie's grin widened and he threw my bag on the floor of the hut for effect. He wasn't joking.

I surveyed my new home. The reed hut had no door, only a door-shaped opening at the front, and the thatched roof was strung with spider webs. There were no windows, only a rectangular opening of latticed reeds at head height to let the air in. There was barely enough room for two skinny camp beds on the sand floor, and now we were about to add another one.

I sat on one of the khaki canvas beds, the tough fabric held a hand's length off the sand by a contraption of thin but surprisingly strong steel poles. An empty beer bottle with a partly peeled off label that said *Mosi Lager* lay on its side beside a bow and a collection of feather-tipped arrows. Men's rumpled T-shirts and towels were draped around the room in a haphazard array. I pushed a cigarette butt under the sand with my toe. An obese tsetse fly settled on my thigh and began to suck. I swiftly squashed it and flicked the mangled bloody remains away with repulsion.

At around sundown a dilapidated, open-topped Land Cruiser with red-haired Mark at the helm coughed and spluttered into camp. In the back sat Peter on top of a bulging hippo carcass that had been shot that day. The two dishevelled figures that tramped unknowingly to their hut's entrance were covered in dried mud and caked hippo blood, with week-old beards, bare feet and ragged clothes that looked as though they hadn't been washed in a while. Seeing their familiar faces made me feel a little better about things.

'Tam!' they both exclaimed simultaneously. 'What are you doing in our room?'

'Your room?' I replied, smiling demurely. 'You mean *my* room?'

Roger had given me some vague orders before I'd left Humani that had consisted of 'make sure the guys up there do some work' and from Anne I'd been told to 'give the place a lady's touch'.

Whether I liked it or not, in Zimbabwe and Zambia, as a woman, it was assumed that I would undertake a certain role. In this case I automatically became the 'Madam of Malambwe' and was responsible for the kitchen. At Humani, when I told the guys that I was going off for a walk in the bush because I was interested in wildlife, they asked me if I was a feminist. This was a place where women were women and men were men. When in Rome, I did as the Romans did, even if it went against the grain.

The khaki canvas tent that was the kitchen resembled something out of an army barracks in *MASH*. A fridge and separate freezer were the only vaguely recognisable kitchen appliances in the entire room. A workbench made of thin mopane logs tied together with bark was visible under a stack of boxes full of miscellaneous tins and cans. Like my sleeping quarters, the floor was soft sand. A cautious poke into some of the unlocked trunks revealed a serviceable first-aid kit including things such as malaria tablets and snake antivenom treatments. An abundance of hairy-legged spiders scattered wildly from under a box of towels and sheets, as thrilled with their sudden discovery as I was.

Beside the kitchen in a meagre enclosure of vertical logs was the cooking area. A gently smoking mopane campfire sat beside a rectangular hole in the ground covered by a scrap of old tin. The latter was used to bake bread by putting a loaf in the hole and some hot coals onto the tin lid. It made some of the tastiest bread I have ever eaten.

I was introduced to the Zambian men who would be my camp staff – Phinius, the cook; Boyd and Silia, the waiters and kitchen helpers; and Douglas, who was responsible for getting water from the river for the camp. Boyd, a five-foot dwarf of a man with a wide grin and intelligent eyes, would come to help me in more ways than I could know.

He demonstrated his impeccable English on the first day when he said, 'I am thinking this... Here is not too much food. You must be making a list, madam.'

'You don't have to call me madam, Boyd. Tammie is fine.'

'Yes, madam.'

After having a good look over the supplies we'd brought with us from Zimbabwe and what was left in camp, Boyd's advice seemed particularly useful. I put together a list of the everyday basic necessities for a camp and showed it to Mushie.

As he read down the list he said aloud, 'No...no...no...no... Vanilla? No chance! Tea biscuits? Maybe...if you're lucky... No...no...'

When he was satisfied, he handed the paper back to me. My jaw dropped – a vulnerable gesture in tsetse country, which occurred only when the bearer was absolutely stunned, stark raving mad or indiscriminately starving. Only a fifth of the items on my list didn't bear Mushie's jagged ink stroke of rejection. I was disbelieving and overwhelmed. How did these men expect me to run a camp without enough food to keep the clients happy?

'This is Zambia, Tam,' Mushie answered with a hapless smile.

Those three words would burn into the cerebral hindquarters of my memory as if branded there with a red-hot iron. 'This is Zambia' – the all-encompassing phrase which was the logical excuse for everything that didn't work, went wrong, wouldn't start, wasn't available and just gave you the screaming horrors. If the generator broke down and everything in the fridges melted, the shower water wasn't hot, there was sand in your sheets, the client was eaten by lions, a spider bit you on the butt on the long-drop or any other such mildly ulcer-rupturing situation, all you had to do was remind yourself that this *was* Zambia where these sorts of things happened as a matter of course, and you immediately felt a lot better.

I hadn't known what my job would entail when I'd volunteered to work in Zambia, but now that I was here, I was determined to make the most of it. At the very least, buying groceries for the entire camp would be an adventure. After all, this was what Sarah had been training me to do. Now I would be running a camp on my own. I wanted to prove that I could do it.

Despite his misgivings about how much we would achieve in a shopping trip at the nearest town, Petauke, Mushie handed me an average-sized wad of one hundred thousand kwatcha the next morning at dawn.

'Remember, Tam, we're on a budget,' he added.

The Bible-thick wad of notes in my bag wasn't worth more than US$200; nonetheless, it was the closest I have ever been to feeling like a millionaire.

With Peter driving, the Land Cruiser lurched along the horrendous track in second gear for the two hours into Petauke. As the track wound into town, we passed women wearing brightly coloured sarongs, carrying heavy loads in baskets on their heads and small, sleeping babies wrapped in cloth on their backs. Petauke, I realised, wasn't really a town. It was the nearest commercial centre for miles, but it was more of a single beansprout than a vegetable garden, if you know what I mean. In reality, Petauke was nothing more than a row of corner stores run largely by turbaned Indians with winning smiles and flashing gold teeth. There was a BP station with fuel that cost more than anyone could afford and a colourful marketplace that teemed with people selling home-grown vegetables like tomatoes and peanuts, cheap imported Zimbabwean cigarettes and American foreign-aid clothing. Of course, there was also a pub, heaving with stocks of *chibuku* and *sadza*, the corn-based staple diet of both Zimbabweans and Zambians.

As we began shopping, the precious wad of notes in my bag disappeared faster than I could have imagined. I passed over one hundred-kwatcha note after another to one smug, golden-toothed Indian after the next.

My presence in the town caused quite a stir. The way people were staring at me suggested that they very rarely saw white women. The way a baby began to cry when he saw me suggested that the children may never have seen a white woman in their lives and probably thought I was a ghost! It is a strange feeling to come from a country in which white people are the majority into one in which they are the minority.

Whenever Peter went into a shop and I was left alone, local teenagers took no time in approaching me with well-practised pained expressions, begging. It always began with, 'Madam, I have a problem.'

In Zimbabwe I'd been living on a commercial farm where everyone had enough to eat and a place to live. Here in this part of Zambia people were desperate for jobs and food and they weren't shy about begging. To them, we must have appeared very wealthy. Nonetheless, it didn't feel right to just hand over money. These people obviously received a lot of foreign aid, but what had handouts achieved apart from providing a short-term solution for their hungry bellies? People were still begging. What they needed, it seemed to me, was self-sufficiency: the skills and training to grow their own food or to get jobs that paid enough to feed their families. But at that point, I really couldn't comprehend the scale of Zambia's problems. All I knew was that I was very lucky to have been born into a family in Australia.

Later I discovered why the people of Zambia were suffering so much more, it seemed, than their southern neighbours. Zambia gained its independence in 1964, eleven years before Zimbabwe, but declining copper prices on which the economy depended, prolonged drought and political corruption sent the country into economic despair and widespread poverty. Things began to improve slowly in the early 1990s due to a shift in the political scene that led to economic reforms, but from what I saw in 1995, they still had a long way to go before the average Zambian rose above the breadline. Everything in Africa takes time. History cannot be hurried and time assumes a different meaning in developing countries.

We wandered through the markets, hounded relentlessly by stallholders. Rich earthy colours – maroon, mandarin and gold – tantalising the senses. The air smelt of peanuts and human sweat.

'Buy! Madam, buy! *Tss!* Madam, look!' women in western T-shirts and traditional, brightly coloured wraps screeched as we passed each stall.

The local produce was displayed on blankets spread on the dirt. Together, the stallholders created an incessant bustle like an overexcited pen of gobbling turkeys. The constant noise seemed to follow us, growing in volume and intensity when we stopped to examine a pile of tomatoes or to ask a price.

Our priority was to buy some eggs. The clients, a retired American couple, had been on safari with Hilton many times before.

'They're not too temperamental,' Hilton had told me. 'Just make sure they get their boiled egg of a morning and they'll be happy.'

After searching the entire town for eggs, we discovered a stall with about six dozen dull brown ones displayed on a cardboard crate. Not wanting to risk running short in camp, I bought most of the shop out. At last, our arms laden with a colourful array of sun-hardy produce including sweet potatoes, onions and even popcorn, Peter and I staggered back to the Cruiser. Unfortunately, Mushie's assessment of what we would and wouldn't be able to buy in Petauke had prove correct. Despite this, we were returning to Malambwe with a lot of what we needed to feed the camp. I was a happy – if unlikely – madam.

It soon became evident to me that the camp cook, Phinius, was, to put it politely, a little bit unreliable. A grey-haired, stringy man, he purported to have worked in a fancy restaurant in Lusaka prior to becoming a safari cook. He proudly produced a tattered menu, complete with unidentifiable stains and a missing chunk of pages in the middle, as evidence of his credentials. A neurotic sort of fellow with a gummy grin, Phinius seemed to bounce around the kitchen with enthusiasm, as if trying to look rampantly busy, but only when I was looking. His English appeared to be adequate, his hearing okay, but there was a major problem with his comprehension. Anything I suggested to him went in one ear, buzzed around for a while like a rabid blowfly and then flew out the other ear. He assured me he could read, so when telling him what to cook and when to serve failed to have any impact, I wrote instructions in simple English into a little notebook to remind him. Other women had been sent from Humani at various stages to work with him but they hadn't lasted. Now I understood why.

'Yes, madam. Yes. Okay, madam,' Phinius said, nodding profusely and grinning with a gappy yellow smile.

Unfortunately the word 'okay' didn't mean that he understood. To Phinius, 'okay' meant that the 'madam' would get off his back so that he could get on with what he wanted to do. Time and time again meals were served late and weren't of the standard expected by high-paying safari clients. When I tried to help, he grew irate and exclaimed, 'No! I will do it!'

Then he started mumbling incoherently about having cooked for hundreds of people at the hotel he'd worked at in Lusaka. My nerves were on edge, as were his. I felt too young and foreign to be telling him what to do, but I was only trying to do my job. If the meals he produced were of a poor quality, it would reflect poorly on all of us.

The other two staff, Boyd and Silia, were a pleasure to work with. In my two months working at Humani, I'd picked up enough from Sarah about how things needed to be done in the camps to ensure there was a high standard of accommodation. Boyd and Silia were keen to learn the correct way to make beds and fold towels and soon were competing for the jobs. I cut up a sarong that I'd bought in Petauke and sewed it into half-a-dozen serviettes. On the first day I showed them how to fold them, but after that, each evening a new creative folded design would appear next to the wine glasses on the table at dinner.

Douglas, whose main job it was to fill the small suspended tank which served as a shower and ensure that there was always water from the river for washing up, spent most of his day sitting idly near the kitchen smoking rough local tobacco in rolled newspaper cigarettes. There was a timeless indolence in the air at Malambwe, a wafting, tobacco-flavoured easiness in the atmosphere which made it the kind of place where nothing ever got done quickly. As Douglas spoke little English, I asked Boyd to translate for me and to ask him to rake the sand around the main dining area. However, my suggestion gained a lot in the translation and the dining area floor was transformed into an artwork of raked strokes and swirls. I almost felt guilty to tread on it!

Phinius and I, however, continued to make each other's blood boil. He didn't like the idea of having a madam around the place because this meant only one thing: work enforced by an interfering woman. One morning I heard a heated argument coming from the direction of the kitchen tent. There was one dominant, hectoring voice, but I didn't recognise it. I wandered into the kitchen wondering who was giving the staff a hard time, but found only the four of them there. As I entered, the room fell silent, except for one voice. It belonged to Phinius.

'What's going on, Boyd?' I exclaimed, my voice overwhelmed by Phinius's vehement tirade. Before Boyd could answer me, Phinius turned and noticed that I was in the room.

'Phinius, the client is sleeping in her room. You must talk quietly,' I reproved.

This only served to provoke an even noisier harangue at me from the obstreperous old bloke, most of which I didn't catch because even on a good day his pronunciation was unintelligible.

I was still not sure what had him so worked up, but I wasn't prepared to be yelled at by anyone. On the verge of telling him to get a grip, and a little scared at what might happen when I did, I was saved by Boyd.

Interrupting Phinius's spiel in a calm but stern voice, he said, 'Mr Banda, this is a safari camp. You should not give your political speeches here.'

Dear five-foot-high Boyd – I almost hugged him! Phinius stared at Boyd malignantly, trying to burn him with his gaze. He stamped one foot dramatically like a temperamental child, then spun on his heels and marched off into the bush in a huff.

Boyd, Silia, Douglas and I burst out laughing, but dared not laugh loud enough for the grumpy old cook to hear us. Boyd explained that Phinius belonged to a particular political party and was intent on converting all of the camp staff to staunch supporters. I had walked in on a political rally!

It seemed out of character to me for the old cook to be so vehement, even if it was about politics. But I discovered later that night that there was more to the story. Hilton hit the roof when he discovered that the bottle of whisky in the camp dining area had reduced by a third in the last twenty-four hours. No one in the camp drank whisky. But it seemed our cook had a taste for it.

Phinius also had a propensity for overusing the most rare and important ingredients. Towards the end of the safari, the unthinkable happened: we ran out of eggs. Hilton's words rang in my ears... *Just make sure they get their boiled egg of a morning and they'll be happy.*

In most other parts of the world, if you found your fridge was short on something, the problem could be easily rectified with a quick trip down to the corner store. From Malambwe Camp, the nearest corner store with a fridge was a nine-hour drive away.

'There is only one egg left,' Boyd announced sombrely.

'One egg? You mean one *carton* of eggs?' I said, thinking Boyd must have been confused with the English.

'No, madam,' Boyd stated decisively, gesturing to the small, lacklustre, lonely egg throned in the otherwise empty cardboard box we'd bought.

Panic caterpillars began to wriggle and squirm inside my stomach. There was no operational vehicle in camp other than the one that was being used daily for hunting, so a trip to Petauke was out of the question. Desperately I radioed Mushie, who was working at Roger's other camp in the Bengweleau Swamps, and asked him what I should do. He sounded irritated and told me in no uncertain terms to 'make a plan'.

That must be Africa's unofficial motto. Nothing ever went unrectifiably wrong and a disaster was never a disaster because you could always 'make a plan'. So, in the true local spirit, I sent one of the camp staff, Steven, to his village with orders to return with as many eggs as he could get. The clients were paying around US$250 per day to be here and the least they could expect was their boiled egg of a morning. Steven returned later that afternoon with a one-litre bottle of golden cooking oil.

Boyd translated for me, 'He says there are no eggs in the village, madam. He says next week the shopkeeper will have eggs.'

It seemed ironic that out here in the middle of the Zambian outback you could buy processed oil but, despite an abundance of chickens, you couldn't buy eggs. The chooks were on strike.

The next day another staff member, George, presented me with a basketful of ladyfinger bananas. They were plump and ripe and utterly delicious. Thinking that the abundance of bananas in George's village might be an indication of what else was on offer, I sent him off that afternoon to see if he could find any eggs. He returned on sundown with a cardboard box. It was filled with rape, two sweet potatoes and, thank the Lord, eight quail-sized eggs.

It seemed that the chickens in Zambia, like some of the humans I'd met, didn't grow very big.

'But it is good to be small,' Boyd insisted, chuckling. 'Short people live longer than tall people.'

'Oh really?'

'Ja, when you go from this place, you are teaching us cooking, so we will remember this short madam.'

Later that night, after dinner, I heard Roger's voice crackle over the long-range radio in Mushie's hut.

'Tell Tam I want her back at Humani. She can come out with you guys at the end of the safari.'

So that was that. Almost as abruptly as I'd gone to Zambia I was leaving. I couldn't help feeling a little sad. I'd enjoyed the challenges of running a camp in a remote wilderness and felt I was just getting a handle on the Zambian way of life.

Despite Peter's propensity for snoring like a tractor, I figured it probably helped keep the hyaenas away from the opening of our hut, and the two boys had become good mates during my short spell at Malambwe Camp. I'd miss Boyd, who'd come to my rescue on so many occasions. Zambia had taught me that even in the most macho, rugged environment, I could still carve a niche for myself and get the job done. It was all about 'making a plan'.

I didn't think that lions hunted during the heat of the day – but I was wrong. Moments after Naomi and I crumpled into the car, chased there by this lioness (*top*), who probably thought we were a threat to her rhino carcass, she and her pride took down a springbok just fifty metres from us (*middle*). Then a male emerged from the long grass and monopolised their kill (*bottom*). With our hearts in our throats, we watched the lions devour the springbok, knowing that it could have been us.

At the age of fifteen, my view of the world was turned on its head when I visited the Save Valley Conservancy with my father, who was on a hunting safari. As an avid animal-lover, I couldn't understand how hunting could play a role in nature conservation. JULIANNE NICHOLS

As a seventeen-year-old English teacher at Humani Primary School, many of my pupils were older than me. After all, I was fresh out of school myself! PHOTOGRAPHER'S NAME UNKNOWN

The Malambwe Camp kitchen in Zambia was like a scene out of *MASH*. (*Left to right*) Silia, Phinius the cook, Douglas and Boyd clear the coals to lift a loaf of bread from the underground oven.

My reed and grass hut at Malambwe Camp didn't offer much protection from the hyaenas and tsetse flies. But it made a surprisingly comfortable home while I lived in Zambia, even if Pete, my hut-mate, snored like a tractor.

At Humani, Jessie the cheetah was very much part of the family and had the run of the Whittalls' house. Here she investigates my freshly baked Anzac biscuits, before deciding that they weren't really her cup of tea.

Exhausted after another unsuccessful outing during her 'training-to-hunt' phase, Jessie is too tired to jump into the back of the Cruiser. Roger was unimpressed with the cheetah's slow progress as they tried to return her to the wild, but Anne's determination later paid off.

Karen Paolillo's house was a utopia set against mopane woodland on the banks of the Turgwe River in Zimbabwe. Her small bedroom, in front of the house, is completely open to the elements. Once, late at night, she watched a leopard walk past her bed as she lay there.

Hippos were the subject of Karen's long-term behavioural research and also part of her family. Here, at sundown, a hippo gapes, quite relaxed in the presence of Karen and myself on the Turgwe's banks.

A greater egret takes flight on the banks of the mighty Kunene River, which runs along the length of the border between Angola and Namibia to the Atlantic Ocean.

Two baby jackals inquisitively watch tourist vehicles driving past from the safety of their den beside the road. As a researcher, there was so much more to see than just my study species. Every day brought with it new sights, smells and adventures.

No matter how many times we got bogged, which was at least three times a day in the wet season at Humani, Ipheas never failed to keep grinning. It was all in a day's work for us.

The roads at Senuko became treacherous as the rainy season bore down on us; traversing the muddy tracks became as much of a challenge as studying the impala. On several occasions we had to walk back to the nearest camp to beg for a tractor to tow out the blue beast.

In the Save Valley Conservancy, common impala are as numerous as flies in the Aussie outback. Locally they are known as Zambezi goats, and most people value the species more as good steak than as quality game viewing. But I discovered that there's more to impala than meets the eye.
ALLAN MATSON

Graham Conier *(right)* and his farm hands spent hours digging through this reeking and maggot-infested elephant carcass, looking for a bullet. When the war vet invasions began in 2000, opportunistic poachers took advantage of the chaos the instability had created. Tragically, the animals in Zimbabwe have paid too high a price for the human race's insatiable hunger for land.

WHERE THE WILD THINGS ARE

Jessie and I did not get off to a good start. First of all, she stuck her furry face into my freshly baked plate of Anzac biscuits uninvited and stole two in quick succession. If that wasn't enough, she then proceeded to spit them out as if they were the most disgusting things she'd ever tasted! Now if that isn't an insult to my modest cooking abilities then I don't know what is. Then again, cheetahs are supposed to be carnivores, so I guess I shouldn't have taken it personally.

Jessie was a teenage cheetah with attitude. She had arrived at Humani into the loving arms of Anne after her mother and brother were killed in a bushfire. Anne had a reputation for successfully raising wild animals, including zebras, leopards and lions. These wounded or orphaned animals became part of the Humani family, so it wasn't unusual to see a half-grown lion lazing on the couch in front of the TV, a zebra foal jumping on the bed, an eland nibbling on the clothes on the washing line, or a leopard cub sleeping on the office floor. The idea was to rehabilitate the animals until they were ready to make their way in the big, wide world. Humani was a foster home for orphaned animals.

When I met Jessie, she was testing out her skill and agility on unsuspecting humans. I was innocently walking in the garden talking to the butterflies when she suddenly appeared in a rush of gold and black spots, pounced on my leg and then disappeared as quickly as she had come. The attack had been so fast that I was barely able to identify my attacker, who was almost fully grown and weighed about thirty-five kilograms.

A few minutes later, as if playing a game of hide-and-seek, she prepared to rush at me again from behind a bush. Luckily this time I saw her coming and was ready for her. She was crouched low, intent on her prey, ready to pounce.

'Jessie! No!' I said firmly, spinning to face her as she prepared to cross the final few metres between us.

Instantly she stopped in her tracks. Now that she'd been discovered, I wasn't any fun as a target any more, so she swished her long, ink-tipped tail into the air in a devil-may-care fashion and sidled up to my feet. Like a temptress, she rolled over onto her side in a pretence of docility. She was a beautiful creature with soft, spotted fur. I reached down to stroke her behind the ear, only to have my hand swiped at with a sharply clawed paw. She always had to have the final say. It was a matter of pride.

The Humani dogs, an assortment of Rhodesian ridgebacks and short-haired pointers, were not very impressed with the latest addition to Anne's foster family. In an ironic twist of fate, now it was the dogs that were running away from the cat! A series of yelps often signalled that Jessie was bored again and someone would have to run out to coax her away from an indignant canine by shoving her 'ball' in front of her. She loved her ball, a sugar sack rolled up and attached to a long piece of rope. Jessie was only playing with the dogs, as she did with the chickens and the ducks.

Once she had caught some unsuspecting feathered waddler she simply let it go because she didn't know what to do with it next.

No one, least of all me, could deny that Jessie had the instinct for the hunt, but she didn't seem to have it for the kill. This was something she'd have to learn and some things are best learned the hard way. Anne announced one day that it was time for Jessie to learn how to hunt and kill like a cheetah. There would be no more chasing of dogs and chickens and Aussie girls for her. The time had come for predator training.

Anne jumped into the driver's seat of the single-cab Hilux and Sarah and I coaxed Jessie into the back with her ball. We held onto her tail with both hands to stop her jumping out of the back and chasing the goats and children in the village. Through dark brown eyes with long lashes she watched the open-mouthed people we passed with growing interest, cheeping excitedly. A cheetah doesn't roar but cheeps like an oversized canary.

We drove beyond the airstrip to a place where we knew there were plenty of impala. Common impala weigh about forty kilograms – just the right size antelope for a cheetah like Jessie. Then we let go of her tail and she leapt from the car without hesitation. She walked off in an unaffected saunter, as if she was the coolest, grooviest chick on the planet and did this silly ol' hunting thing every day of the week.

Had she grown up in the bush with her biological mother she would have learned how to hunt and kill as a matter of survival. However, our Jess was accustomed to a higher standard of living, with a juicy impala leg delivered to her doorstep every evening, free of charge. Hunting, to Jessie, was something to be done in playtime. To our dismay, although she belonged to a species that was the fastest short-distance runner in the animal kingdom, she would get puffed on the five-minute walk between house and office. This didn't bode well for her future as a wild cheetah.

Undeterred by any of this, Jessie clearly considered herself a superior being. Being a lady cat of great culture and immense intelligence, for her very first stalk in the wild she decided she was far too advanced to waste her time on a small antelope like an impala. She wanted a kudu, which is about twice the size.

Unfortunately, even a cheetah's ego can be deflated. We watched her stalk and then run up to an unsuspecting herd of kudu cows, and with great skill too. The kudu had continued to browse on an acacia bush, their grey and brown coats dappled in the afternoon light, only becoming aware of her when she began to sprint. But Jessie made a slight error of judgement. She would carry the mark of this first defeat for the rest of her life: a deep scar under her eye. Jessie came at the kudu from the rear end, did her best to jump onto its back, only to be kicked in the face with perfect accuracy by a sharp hoof. She was lucky not to lose her eye as she stumbled around in a drunken circle while the kudu pranced away. Poor Jessie. It was a lesson learned the hard way, but it served its purpose. After that, she lowered her sights to hunting impala and other smaller antelope. Rarely did she catch anything on our afternoon hunting forays, but she was learning.

Eventually Jessie wandered off into the bush and never returned. No one saw her again and, to our sorrow, we all presumed she'd either died or moved away. Reintroducing captive-bred animals to the wild is extremely difficult because ultimately human care, no matter how well intentioned and devoted, is a poor substitute for a real mother and a wild upbringing. Only the strong survive in the wilderness and, sadly, animals bred in captivity are usually at a significant disadvantage when released into the big wide world of competitors, predators and prey.

Several years later, while driving around Humani, I caught a glimpse of a rare sight: a mother cheetah with two cubs. The mother sat tall and proud, watching me intently over the grass, while the two cubs gambolled around her, occasionally poking their adorable heads up to look at me inquisitively. It was strange that the cheetah didn't run away immediately, especially with two vulnerable cubs. Was it recognition in her eyes that I could see? I pulled out my binoculars and took a good look at her. There, to my delight, was a scar under the mother cheetah's eye.

Cheeky cheetahs are one thing, but wild elephants are another ball game altogether. Most Africans I know don't need to *learn* to respect elephants. A healthy dose of fear is so deeply ingrained in the people who live in rural villages, where elephants threaten their lives and demolish their precious crops, that no one questions their authority. In contrast, most westerners I know adore elephants and see them as the inspiring gentle giants they are so often portrayed as in television documentaries. There is validity in both of these views.

Elephants really are extraordinary creatures with intricate social bonds, a poignant awareness of death and apparently the ability to express grief, as well as an undeniable ancient wisdom. At the age of eighteen I erred towards the bunny-hugger view of these proboscidean monoliths. When walking in the bush I was always on the lookout for them, but in truth I never understood the very real danger an elephant poses to a human on foot until I almost walked into one myself.

My friend Brenda, who was guiding a couple of dear old English ladies on a horse-riding safari, invited me to join them at Sabi Camp for the evening. Sarah didn't have any work for me to do that afternoon, so I decided to take the opportunity to go for a bush walk in the direction of the camp. Brenda said she would pick me up on her way home with the ladies, about an hour later, before it got dark.

There is nothing I love more than walking in the bush, with the blood pumping through my veins, the fresh, sweet aroma of the land in my nostrils, trees and animals for company. Energy surges through me like a living force and I feel powerfully connected to the earth.

I'd been walking for an hour when I reached the rhino *boma*, a tall mopane enclosure close to the hippo dam. The sun was beginning to sink below the horizon, painting the sky a glowing orange. Brenda should have been here by now. I hoped she hadn't forgotten me. A tiny tickle of self-preserving fear nagged at my unconscious. Something told me to turn around and start walking back. Whatever sixth sense that was probably saved my life, although I didn't know it then. I was still enjoying myself. A happy-go-lucky warthog strutted jovially along in front of me, its tail erect and its chest thrust out in self-important pride. My heart leapt involuntarily as a nervous duiker sprung from the undergrowth beside me and bolted away into the twilight. A twinge of edginess pricked my senses again and I began to walk a little faster.

To my left, the electric fence surrounding the sugar cane stretched for kilometres, with wires too closely placed to allow a human to pass through. To my right was bush, the wild savanna that Sarah once described as 'the deepest, darkest Sabi'. Tall, elegant ilala palms towered above a sheltering canopy of flat-topped acacias like sentinels guarding the heart of the forest. A blanket of dense low-lying bush with finger-length, gouging thorns covered the cracked earth beneath.

Time began to slow down. It was a quarter past six and quite dark, with a mere sliver of a moon for light. I began to pick up my pace to an adrenaline-pumped jog. This wasn't fun any more. In retrospect, with the benefit of experience, I now know that jogging was the wrong thing to do. If there had been any predators around, I would have been a much less appealing target if I'd been walking rather than running.

When I finally reached the mealie fields, I decided to squeeze through an opening in the barbed-wire gate where it was not electrified and crouch there until Brenda came along. For some bizarre reason I felt safer there than I did walking along the road, as if the upright, green mealie plants were sheltering me from the night's predators. I sat there, deathly quiet, my sense of hearing suddenly acute. Even the pounding of my heart in my chest seemed to magnify and thump like a jackhammer in a monastery. Brenda will be here soon, I reassured myself. Time continued to tick by at a painfully slow rate. Fifteen minutes later there was still no sign of an approaching vehicle.

I was a good thirty minutes walk from the house, which doesn't sound like much. But I'd seen a lot of fresh elephant spoor earlier and I knew they'd been around the house in the last week. There was always the possibility of walking into a leopard, too. Our poor night vision makes us humans so much more vulnerable in the dark when the senses of the predators are far superior. We tragic primates of the human variety don't have horns or canine teeth to defend ourselves with, which makes us ridiculously vulnerable in an environment where things can eat you. All we have is a large and well-developed brain: our smarts are our only defence. My own overheated brain ticked away furiously. Should I stay in the safety of the mealie fields, knowing that someone would eventually come looking for me, or should I make a run for home? The first idea seemed like a better option – at least for the next five minutes.

That was when I heard footsteps.

It was a large animal, not as big as a fully grown elephant, but certainly bigger than an impala, judging by the sound of the footsteps. The blurry shape in the darkness was coming towards my hiding place in the mealies where I crouched like a hunted animal. The rhythmical patter of feet on dirt had taken on a thundering volume. That, combined with my thumping heart and heavy breathing, made for a terrifying cacophony of sound in the deathly still darkness.

If I'd known a bit more about African animals I would have known that only one creature makes that much noise in the bush. Elephants and lions come so quietly you wouldn't even know if one was standing behind you. But any logic had long since forfeited its place in my brain to the clouded, nonsensical clutches of panic. All I could think was that this wasn't the Aussie outback and here I was as vulnerable to Darwin's theory of survival of the fittest as the average meal-sized impala. I could feel my chest tightening by the second as the footsteps grew louder. *Damn it, where was Brenda?* I squinted, forcing my eyes to sharpen and focus on the approaching fuzzy shape. At last, I started to make out a figure in the darkness.

The woman carried a sleeping baby strapped to her back with a blanket, an umbrella in her hand and a variety of *sadza* pots and pans in the other. She was probably heading home after a day of shopping in the Tribal Trust Lands near Humani. The woman was walking quickly and decisively, despite her load. She passed me and carried on until her shape became a blur again in the dark night. So overwhelmed by a feeling of relief was I that it took me a little while to realise that the woman offered me a great opportunity. If, as it appeared, Brenda had forgotten to pick me up, it might be my best chance of getting back to the house.

Impulsively, I sprang out from my hidey-hole under the fence and jogged up behind her. Sensing a presence behind her, she increased her pace so that I had to speed up to catch her. Afraid that I might scare her and blow my escort, I let her know I was not a threat.

'*Manheru!* Good evening!' I called out. '*Unoenda kupi?* Where are you going?'

The woman didn't slow down and appeared to be as much on edge as I was. She answered softly as I began walking beside her, '*Oo-mani,*' in the local pronunciation of Humani. If she hadn't been just as terrified as I was, I would have hugged her.

Neither of us spoke again as we hot-stepped it along the thornbush-lined road in the darkness, but there was enormous comfort simply in the presence of a fellow human. We marched in silence for about twenty minutes. Both of us needed our senses to be on full alert – we weren't out of danger yet. The smell of fresh elephant dung filled the air like composting vegetables. For all we knew we could have been walking right beside one. I tried not to think about it and focused instead on the long road ahead.

Finally, two white beacons of an approaching vehicle's lights, one characteristically brighter than the other, signified that Brenda was coming at last. For the first time the woman smiled. We were less than ten minutes walk from her village and the tension was diminishing with every step.

'*Ndatenda,*' I thanked her.

Thank you seemed vastly inadequate under the circumstances, but the woman's eyes told me she understood. The Land Cruiser blinded our night vision and the woman disappeared into the night in the opposite direction. I had just met a guardian angel.

'Hey, Aussie girl. Sorry man, we got a bit carried away with the wine...' Brenda explained sheepishly.

I was too pleased to see her to be angry. As we carried on towards the camp in the rattly old Cruiser, a mountain of an elephant bull emerged on the road in front of us. He was standing exactly where I'd stopped and turned around an hour earlier. The dust from the Cruiser's passage seemed to swathe him in wraiths of mist in the dim headlights. He was simultaneously magnificent and terrifying. The enormous creature moved silently off the road like a ghost and vanished into the surrounding bush, allowing us to risk passage.

The road past the hippo dam was a narrow, elevated track over the dam wall. Water and bobbing lily pads with dusky pink flowers lapped within feet of the vehicle tracks, where, earlier in the day, diminutive jacanas had danced. There was just enough room for one vehicle to pass.

Suddenly the most chilling sound I had ever heard cut the silence of the night. As we rumbled along in the Cruiser with virtually no brakes, very little suspension, no roof or doors, we were trapped between the deep water on our right and a roaring, charging elephant on our left. The bull exploded out of the bush, reared around on his hind legs and trumpeted furiously. I barely registered the thought of what could happen if we were pushed into the dam, which was full of crocodiles and hippos. Perhaps the bull was in musth, an aggravated, hormonal condition during which male elephants seek mates and fight other bulls, or perhaps our vehicle had invaded his personal space. Whatever the reason, he was furious and we had to get out of there fast.

'Go! Go!' one of the grannies screamed, bouncing up and down in her seat.

I couldn't speak. Brenda's eyes widened as she pressed her foot down hard onto the accelerator. I clutched my seat so tightly my fingernails threatened to jab holes in the leather. The Cruiser surged over the dam wall as the elephant tore towards us, and Brenda didn't slow down as we bounced down the other side. The bull chased after us for about fifty metres, but Brenda drove like a bat out of hell until, still panting with fear, we pulled in at the familiar thatched dwellings of Sabi Camp.

Tucked in bed later that night I humbly contemplated my own mortality. On days like this one, Africa's lure was stronger than ever. I wondered why. It was more than my growing friendship with the Whittalls and the feeling of being one of the family. It went beyond Humani too, although I loved the farm and felt at home there. My plans hadn't taken any shape yet, but I knew that I had to find a way to stay in Africa. Its people and wildlife had hypnotised me. I felt alive there in a way I'd never felt in Australia's cities.

The paraffin lamp flickered, creating weird, amoebic shapes on the gently undulating canvas walls of my tent. I listened to the mellow hum of night insects and the distant sound of drums and singing from the village on the other side of the Sabi River. I had learned a lot that day. It was good to be alive in Africa.

***

The next morning I splashed my face with water from a plastic bowl on the waist-high mopane stand at the front of my tent. The water was cold but refreshing. At eight o'clock, the mid-August morning was already warming up.

I could hear the rhythmic whoosh of the rake on the sand floor of the open dining area where Mac, the waiter, was cleaning up. Jairos, the camp cook, was refilling jars of jam in the reed-and-thatch kitchen when I wandered in in search of a cup of coffee.

'Morning, Jairos,' I said.

He looked up and grinned effusively. 'Morning, Tammie!'

There was something about Jairos that always put a smile on my face. At about thirty he was one of the youngest cooks at Humani and always full of energy. He was so keen to please that often he ended up making a mess of things by trying too hard, which resulted in him having a reputation as the local comedian. He was a constant source of amusement.

I asked him once whether I'd been given an African name. Earlier, one of the Humani bachelors had proclaimed that my name was *Nyamtuta*, but I didn't believe him.

'*Nyamtuta*. Dung beetle,' Mark had elaborated with a conspiratorial wink at Peter, 'because you spend your days organising shit.'

The dung beetle is one of the most industrious workaholics in the animal kingdom, so it wasn't altogether an insult.

Jairos's shiny black face lit up with a beaming grin and he chuckled to himself. 'Ja, *Nyamtuta*. This is your name.'

Jairos's springy mat of curly black hair was capped with a different hat every time I saw him. On one occasion, when I complimented the particular hat he wore that day, he beamed and said, 'Ah Tammie, you cannot have this one.'

I laughed as he dashed off.

'You cannot have this one, but *this* one you can have.' He grinned and presented me with a handmade palm hat from behind his back.

'Oh Jairos, *ndatenda*! Where did you get it?' I exclaimed, touched by his generosity.

He leaned over towards me as if to reveal a great secret.

'I make a deal,' he whispered, smiling cheekily and winking.

Knowing that deals or bribery or blackmail or whatever you want to call it are a normal part of life in Zimbabwe, I thought no more of it.

Raucous vervet monkeys were frolicking and screeching in the trees near camp. I decided to go for a stroll with my camera to see if I could get a closer look at them. I walked along the dirt vehicle track, looking at the spoor of all manner of antelopes, from the secretive nyala to the diminutive duiker. Baboon footprints, like human hands, pockmarked the track.

A baboon barked close by, the sharp warning call startling me.

'Bah!' he shouted, 'Bah-hoo!'

My feet made prints in the dirt over the crescent-shaped indents of antelope hooves. The tracks of a solitary bushbuck had been overlain by the more recent visitation of a herd of rufous impala, evidence of the abundant and varied life forms that roamed along the riverbed.

Everything in the bush seemed to be connected into a single living, breathing entity, from the largest of trees to the smallest of insects, all were interconnected and interdependent. For example, elephant dung. The elephants fed on the acacia trees and dispersed their seeds by defecating further afield in the bush. Dung beetles decomposed the dung while feeding on it, while the seeds of new acacia trees germinated in the nutrient-rich, broken-down dung, continuing the cycle.

When I walked in the bush I was absorbed by thoughts about the web of life and how we, as humans, fitted into it. We take from mother earth more than we need to survive – water, food, soil, sun – but what are we giving back to it?

Suddenly my heart skipped a beat. I detected a distinct movement in the grassy gully below. I froze. After last night's encounter with the elephant, I wasn't taking any chances. I stood still, watching intently, utilising all my senses.

I breathed a sigh of relief when I saw that it was only a group of women, striding elegantly through an ocean of long grass and talking animatedly among themselves. I was on the verge of greeting them amicably in Shona when one glanced up and saw me. With a look of horror on her face, she dropped her bundle of what looked like grass or reeds and called out urgently to the others. Like a herd of impala alarmed by a predator, they fled in a babble of panic-stricken warning cries and a rustling of grass.

I felt an all-too-familiar dread sinking into my chest. Had something crept up behind me and scared the women away? I pirouetted on my heels, expecting the worst. But there was nothing there save the wind in the thorn trees.

I couldn't understand it. Then I recalled that Sarah had mentioned something about villagers from the Tribal Trust Lands coming onto Humani to steal palms used to make hats. Roger allowed them to take palms, but only with his permission. He had to monitor his trees as well as his animals to ensure that all use of natural resources on the farm was sustainable. Judging by the brisk response my presence had generated, I gathered that I had been witnessing something that wasn't one hundred percent legal.

I jogged back to camp, less than a kilometre away, for a second opinion.

'Jairos! Mac!' I called.

Both of the men ran out with alarmed frowns to meet me on the road.

'There are women in the bush with ilala palms,' I began.

Before I could go on, Jairos exclaimed, 'Poachers!' and he was off down the road like a shot.

'*Uya kuno*! Come!' he called to Mac and me as he ran, almost overbalancing and tripping over his own feet.

Mac was considerably less enthusiastic and strolled reluctantly behind him.

Thinking that the men would simply tell the women to leave if they didn't have permission, I returned to the kitchen and left them to it. After the initial excitement had faded, I began to hope that Jairos and Mac *didn't* catch the women. I understood that there was a principle to uphold and an example to set by reprimanding the women if they were trespassing and stealing, but a poacher of ilala palms was hardly your average automatic-weapon-firing rhino slaughterer. Then again, all life is sacred – plants included. Where does one draw the line?

Jairos returned a half-hour later, grinning from ear to ear and bursting with excitement.

'Tammie! We have caught them!' he panted.

I followed Jairos down into the grassy gully where I'd last seen the women.

'One of them, she wants to cut me with her knife!' Jairos explained, rolling up his sleeves and taking on the stance of a boxer with clenched fists. 'I say, "Ah ho! I will fight you!"'

I nodded and tried not to laugh at his bravado.

'Then one, she says she will pay me. I say *no*!' Jairos went on with exaggerated indignation, as if she had insulted his pride by even considering that he would accept a bribe. 'I do not want money! I do not do that!'

Mac was standing with his arms crossed, a bemused expression on his face, as we approached a shaded sandy clearing where the women were sitting. Jairos launched into an apologetic spiel about only six women having been caught as three of them had cleared the fence into the communal lands before he could catch them. The 'poachers' ranged from teenagers to grandmothers. They sat on the sand with their legs crossed, chatting and giggling among themselves. Jairos passed me three of their knives, handmade with wooden handles and blunt steel blades.

'Take them to camp,' I said, not really knowing what to do but thinking that I should try to look vaguely authoritative.

Jairos seemed to think this was the right thing to do. The women rose lazily and, without any show of defiance, brushed the sand from their bottoms and followed us back to Sabi Camp.

But where did all Jairos's hats come from, I wondered. They were made of ilala palms, after all. I couldn't help wondering if this wasn't a cover-up for his business activities. Back at camp, one of the women asked to go to the toilet and never came back. Oldest trick in the book, I thought, feeling foolish. We gave them a drink of water and they sat patiently waiting for us to decide what to do.

When Brenda returned to camp a couple of hours later, she giggled at my predicament, then said that we should drive them to the house to be dealt with by Roger. By now I was beginning to wonder how I'd got us into this mess and was starting to feel guilty. Thankfully, the women got off very lightly. Roger confiscated their knives and told them to walk back to the communal lands, a distance of about twenty kilometres. By local standards this was a very light punishment – a twenty kilometre walk is nothing to most rural Africans, many of whom walk that kind of distance to get water every day.

Jairos remained extremely proud of himself nonetheless. Last I heard he was looking into a career in anti-poaching.

Capturing palm poachers was one thing, but I felt considerably less threatened by human predators than I did by the animal variety. Animals are predictable to some extent if you can read their behaviour and know the warnings signs. But you can't control the environment, and unexpected things always seem to happen at the least opportune times.

I was helping Sarah with some Dutch clients on a photographic safari when she radioed to tell us that they had a close-up sighting of the secretive black rhino. The head game-scout, John, was leaning on an old cattle fence, chewing on a piece of grass and chatting to a couple of men in the anti-poaching team when we pulled up.

The rhino was asleep, not five minutes walk from where we were standing, John said. He told us to follow him so we could take a closer look on foot.

John ritually thumbed dirt from a handful in his palm, checking the wind constantly as we walked. He made an attention-grabbing *tsss* sound every time he needed us to move direction to ensure the wind was in our favour at all times.

Rhinos have fairly poor eyesight, but they pick up any sign of movement and their hearing is excellent. In other words, if that tonne of pure, concrete-skinned muscle power was by chance to look in our direction, any itches could wait. If charged, the only thing to do is to get behind the nearest tree and try not to move. Any movement is a dead giveaway to a rhino, so even if you are close enough to feel his nostrils snorting warm, wet torrents of rhino breath into your trembling face, you should stay still. Usually the rhino will back off, or so I was told.

This particular rhino looked as though he was enjoying pleasant dreams. To an untrained eye he could easily have been an oversized grey boulder. It never ceased to amaze me how even the largest of Africa's animals were so hard to see in the bush until your eyes were trained to see them. One of the Dutch clients beside me was taking photos like they were going out of fashion. It was incredible to stand and watch this primeval creature peacefully sleeping in his natural environment. Such moments are rare and don't last long – as I was about to find out.

Shockingly, the rhino stood up in the fastest display of getting out of bed that I've witnessed. He stood there militantly, facing us head on, his stance firm and on guard.

'Freeze!' I hissed to the woman beside me, who was still snapping away on her camera.

Luckily, she obeyed instantly, a little taken aback at the seriousness of my tone.

Any moment now, I thought in horror, he's going to charge. The rhino hadn't moved an inch, statuesque but for the flicking of his disproportionately tiny ears. He couldn't see us, but he knew we were there. No one had been stupid enough to make any noise or move – yet.

Just then the camera of the woman beside me began to automatically rewind its film. In the disconcerting quiet of the bush, the whirring noise was deafening. In my mind I imagined a judge striking a gavel in a silent courtroom and sentencing death. I hadn't moved since the rhino had risen, so my neck was still twisted with one eye on him and one on the woman with the camera.

Her eyes registered her desperate panic and she whispered urgently, 'What do I do? I can't stop it!'

'Just freeze,' I said, barely moving my lips. 'He can't see you if you don't mo–'

My heart thumped and I choked on the last word. I surveyed my surroundings by swivelling my eyes like a chameleon, looking for a nearby tree to climb that didn't have thorns the length of my little finger. I was sure that at any second the rhino was going to charge. He had every right to – we'd woken him up from what had clearly been a very pleasant dream.

The rhino snorted. I jumped inwardly. Then he spun on his hindquarters and trooped off into the thornbush. There was an arrogant wiggle to his bottom as he trotted away.

This small adventure, which threatened to give me an ulcer at the age of eighteen, proved my father's theory. Dad always said you know you're having an adventure when you're not enjoying yourself any more. If this is true, then I've had no shortage of adventures in Africa. In fact, I've developed a taste for them.

Later that year I left Zimbabwe destined for university and the start of a deferred law degree. But I knew full well that my future lay in wildlife conservation in Africa, not as a hotshot lawyer in a big city. I was driven by what I'd seen of Zimbabwe's poverty, its magnificent wild animals and the feeling of freedom I'd found in its bush. Its problems were complex and overwhelming, but this only made me more determined to help find solutions. I didn't know how I was going to play a role, and some of my friends and family told me that it was an impossible dream, but I had fallen in love with what had once been called the dark continent, which to me was the land of light. I pined for it back in the relative ease of life in Australia. I couldn't explain why, but there was an inevitability to the fact that I would return. I just couldn't get the place out of my system.

THE HIPPO LADY AND THE WILD DOG MAN

Water cascaded in glorious torrents on its course downstream, splashing over the smooth rocks of the mighty Turgwe River. There is something deeply meditative about being close to running water. The constant flow of a river is like balm for the spirit, as if the sound of water alone is enough to cleanse the soul. I breathed in deeply, trying to take in some of its energy, inhaling the spirit of Africa.

I was in paradise. This utopia was Hippo Haven, the home of my friend Karen Paolillo. A large thatched roof arched over the unique house with its romantic architecture, curving so low in places that tall people had to duck to enter. Thick mopane poles locked into rock walls supported the roof, and huge windows made entirely of mosquito mesh let the light and breeze in. Everywhere were books, photos of hippos and other animals, overflowing teacups and ashtrays full of cigarette butts.

Karen's bedroom was a separate smaller building, about ten metres away from the house, consisting of just two reed walls and a thatch roof. Her queen-sized bed was inches from the edge of the riverbank, where the earth dropped off sharply into rushing waterfalls, intricate rock pools and swirling rapids. It was a completely open-air bedroom, like camping out in luxury. One night Karen heard a cough and woke to see the sinuous form of a leopard padding silently past her bed.

I was sitting on a comfortable log on Karen's green lawn, my chin in my palms, watching with delight the playful primates scampering, fighting and frolicking around me. With inquisitive eyes in fluffy black and white faces, the monkeys inspected me with equal curiosity. While I thought how much their inquisitive, oval eyes resembled a human's, perhaps they were thinking that mine looked like a monkey's. Who knew what went on in the minds of animals? At the age of twenty, I had a hard enough time understanding other humans.

Only a thin stretch of riverine jungle, resplendent in its verdant finery of vines and broad leaves separated Karen's house from the river. An array of bird life, monkeys and baboons lived on Karen's doorstep, not to mention a semi-tame warthog, by the name of Arthur, who visited her lawn on a daily basis. The deep, throaty grunts of hippos punctuated the constant chatter of birds and monkeys throughout the day. It was a magical place – peaceful but never quiet.

The water below Karen's house looked so inviting, but of course to swim in it would have been a death wish. This patch of the river was home to a group of hippos, Karen's study population, all of whom this extraordinary woman knew by name and personality. According to the statistics, hippos kill more people in Africa than any other animal, except for mosquitoes of course. Crocodiles sunbaked on the Turgwe's exposed sandbanks and sank ominously into the murky brown depths, waiting for an unsuspecting victim to wander too close to the water's edge. There would be no swimming here.

Guineafowls squawked from the trees near the house. Strangely, the raucous cacophony of the flock reminded me of my home in Australia. Since my parents had moved from Townsville to a cattle farm on the Darling Downs in south-east Queensland, they'd bought a few guineafowls that roosted in the jacaranda tree beside their bedroom. Secretly I think my mum's ulterior motive in buying the African birds was a hope that they would be enough of a reminder of Zimbabwe to stop me going back again. She was trying to create a little Africa for her daughter on the Darling Downs. Unfortunately for her, the plan backfired. The birds were the size of chickens, with voices to match any rooster, and they made their presence known each day when they decided it was dawn, which was usually around three in the morning. My parents grew to hate their hideously early wake-up calls and their presence only made me long even more to see them in their natural environment.

The weather was *guti* as the locals say – overcast. Behind Karen's house, the *kopjes* were shrouded in low cloud. It had been this way for a few days, frustrating both Karen and me. In overcast weather the hippos moved away from this section of river and were less likely to come up onto a sandbank and sunbake when it was cold. This made observations of their behaviour virtually impossible.

I was back at Humani in my university holidays to learn about hippos from a lady who knew more about these relatives of Porky Pig than most. After all, she had spent the best part of twenty years studying the hippos at Humani. In her late forties, she had hippo children, not human ones. It was as though she had been accepted as a member of their pod, which was an amazing thing, considering that female hippos weigh in at close to one and a half tonnes and males up to three tonnes!

Karen had persuaded Roger Whittall to let her build her home on the banks of the Turgwe. In return, he sent his safari clients to visit her and she took them down to the riverbank on foot to see the hippos. This was a truly amazing experience for any visitor, to see wild hippos so relaxed, sunbaking, snorting and bombing into the water, from only thirty metres away. She had created the Turgwe Hippo Trust, which survived largely on client donations.

Locally, Karen was known for being a little bit eccentric. The game farmers who were her nearest neighbours could not understand why an English-born woman would choose to devote her life to the conservation of a bunch of hippos. She disliked cooking for herself and, as I discovered, she survived on not much more than mashed potatoes and baked beans. I'm quite sure that the monkeys ate better than she did.

When I asked her whether she ever became lonely, Karen simply laughed. With the menagerie of hippos, monkeys, warthogs, cats, goats and other animals living with her, how could she ever be lonely? On top of all her feathered and furry friends, Karen was married to a French geologist, Jean-Roger, who lived at Hippo Haven when he wasn't involved in geological projects throughout southern Africa.

Jean had picked me up from the airport. I'd arrived with the strange mixture of excitement and exhaustion that always follows the horrendous twenty-two hours of travelling from Brisbane to Harare. With jetlag, everything feels like an hallucination. A fat, surly customs officer stamped my passport into Zimbabwe and the look in his eye told me to piss off before he changed his mind. In the waiting area I spotted Jean, a wild and woolly Frenchman with a stubbly jawline and a mop of curly, black hair.

As we drove out of the smouldering city in his rusty *bakkie*, Jean seemed more edgy than I was in the throngs of people. A traffic light, or robot as they are called in Zimbabwe, turned red all of a sudden and he slammed on brakes to pull up just over the white line. He was blinking hard and sweat dripped down his forehead into his eyes. In between blinks, he explained to me that he was working for Rio Tinto, looking for diamonds in the highveld north of Harare. Suddenly he started swearing and cursing, clutching at his groin and jumping up and down in his seat. My eyes widened. Luckily the lights were still red. I could smell something burning. Jean was still swearing and grabbing at his pants.

'What's wrong?' I asked nervously.

The lights turned green and cars behind us began to beep their horns with annoyance.

In a perfectly calm voice, unaffected by the hooting cars, Jean turned to me and smiled, 'Fucking cigarette fell in my lap.'

I looked down at his pants momentarily and noticed that there was a singed hole in them. Hurriedly I averted my eyes, then burst out laughing.

Jean began to relax as we embarked on the drive south from Harare to Humani. With a doctorate in geological engineering, the man had a mind to be reckoned with. In the next five hours of driving we solved all the problems of the world and a few more, from human population overloading to community-based conservation of elephants. It was 1997 and I was in my second year of a degree in environmental science. Despite not having any of the right qualifications for a science degree and dropping out of chemistry in Grade Eleven, I'd pulled out of the law degree before I'd even started, determined to do a course of study that would lead me back to the African bush. I'd read up on the scientific theories behind African conservation and I reckoned I knew it all. Meanwhile, Jean had an opinion on everything and a rational argument to back it all up. By the time we arrived at Humani I was beginning to wonder if anything I'd learned about wildlife conservation at university had any relevance in the real world. But that was why I was here, after all, to gain some practical experience from wildlife researchers working in Africa.

I'd spent many long hours picking asparagus at the neighbouring farm to my parents' to pay for this trip back to Humani. This time, as an undergraduate science student, I planned to volunteer for any wildlife conservation research projects that would have me, starting with Karen's.

'You know, Tam, you're my first ever proper volunteer,' Karen told me a little nervously when I arrived at Hippo Haven, 'and the only reason I'm having you is because you're a friend.'

Blessed with a fabulous size ten figure, a glowing tan and a youthful, vibrant personality, Karen could have been my older sister. I felt as though I had found a kindred spirit. When I'd been working for Roger in Humani's safari operation, I'd loved taking clients up to see her and the hippos and always jumped at the opportunity for a chat and a cup of tea. Karen intrigued me.

As well as her solitary lifestyle and utter dedication to conservation, she was also a deeply spiritual being. She believed in destiny and the power of an individual to make a difference, and she lived her beliefs. She always gave me a different perspective to balance my newfound rational, scientific approach to life.

She was a self-confessed neurotic and a chain smoker; so much energy seemed to exude from her, but she wasn't altogether comfortable around people. It wouldn't be until many years later that I would come to appreciate the way one feels after being isolated from people for a while. Nonetheless, she was determined that I would learn as much about hippos as possible in my two weeks with her. After I'd given a fundraising speech at my college at the University of Queensland in the lead-up to this trip, the students had raised several hundred dollars towards her Turgwe Hippo Trust. Karen was overwhelmingly grateful. Despite my own doubts, she continually told me that I would achieve my dream to work in wildlife conservation in Africa. She told me so often that I began to believe her.

'Tam, you're not on holiday, you know,' she would insist after our seventeenth cup of tea and her twenty-seventh ciggie. 'You're here to learn about animal behaviour.'

On our first morning with the hippos, Karen introduced me to her family. We parked her rackety old Peugeot several hundred metres back from the riverbank and walked from there. The deep resonation of a ground hornbill's calls thumped like a drum not far away. The riverbank was steep and sandy and we grabbed onto the trunks of slender trees as we slid down it on our heels. Karen pointed out a pile of fresh hippo dung that had been deposited the night before. Hippos are herbivores, she explained. They spend their evenings grazing on land near the river when it is cooler and return to the water in the day. They don't have sweat glands, so staying cool is a matter of survival.

As we came closer to the water Karen began calling out in a calm, kind voice, 'Booooobby! Hello Bob! Heeeeyyyy Cheeky! Hey guys!'

She was greeting old friends and they knew her voice. If anyone else had called out to a pod of hippos so close to the water it's very likely it would have provoked a full-on charge. Hippos don't like having their privacy invaded, which is understandable. Karen, however, had habituated these hippos to the point where they no longer saw her as a threat and they accepted other people as long as Karen was with them.

The chocolate-coloured water tumbled down the Turgwe River with unfailing fervour. Reeds bobbed frantically on its banks, salsa dancing in the flow. And there, in the middle of the swirling mass of water, were several faces, staring at us. Their butterfly-like ears flicked a shower of water drops off and focused like swivelling radar on the two humans stepping quietly along a game trail on the bank. Massive flared nostrils squirted torrents of murky water into the air in a startling exhalation of sound. Small pig-like eyes assessed us for a few minutes. One of them yawned widely, displaying ominous, ivory-like teeth.

Karen was still talking to them, 'Hi guys. Howzit, Blackface?'

Her voice was gentle and friendly, a reminder that she meant them no harm. We found a comfortable place to sit on the sand, barely concealed behind some reeds and long grass, and then our work began.

'You see how that one there has mostly pink around her eyes? That's Cheeky, one of Bob's daughters. Bob's the main man, Tam. He's the dominant male. No one messes with him.'

Karen showed me how to recognise each individual in the pod of twelve by their distinct facial characteristics. Marks or coloration around the eyes were useful for indication, as well as nicks in the ears and facial warts. Blackface, one of the older females and the mother of Cheeky, had an entirely black face. That she was easily recognisable was a bonus, Karen said, because Blackface had a habit of charging for no apparent reason and it was good to know when she was around. Blackface, it seemed, was in a permanent state of PMS.

Karen made notes on the hippos' behaviour while we sat and observed them for a couple of hours each day. Often the hippos would walk up onto a favourite sandbank while we were there, lugging their heavy bodies slowly out of the water and into the sun. It looked like a lot of effort on such short, tubby legs. Resting heavy heads on each other's rumps of blubber, they would lie for hours on end, absorbing the heat of the sun like tubby tourists on the Gold Coast. I was surprised to see that often the hippos shared their stretch of sand amicably with huge sun-baking crocodiles. Although crocs are known to kill baby hippos that roam too far from their mothers, the two species seemed to coexist quite happily at Hippo Haven, sharing both terrestrial and aquatic environments. Karen had even witnessed the extraordinary sight of hippos grooming crocodiles on land, perhaps deriving nutrients from their scaly skin.

Bob was Karen's favourite. He made his dominance clear to any potential rivals by marking his territory regularly. He would flick his muscular tail hurriedly while squirting chunky dung against reeds in the water or trees on land. Reeking of his unique hormones, these marking places advertised his seniority to other males in no uncertain terms.

I watched with fascination as an adult female, Lace, groomed her small son by licking the skin on his shoulder for several minutes. Hippos don't have ticks to be removed, which is thought to be a major function of grooming in other species, so Karen felt that this behaviour reflected the need for close mother–young bonds. Hippos are social creatures, living in close-knit family groups. It was hard for me to think of them as the most dangerous mammal in Africa as I watched them snoozing lazily on the sandbank, grooming each other and tolerating a reptile that could kill one of their young.

The hippos began spending most of their time on the other side of the river, which was making observations of their behaviour difficult. Karen suggested we construct a small bridge and cross over to have a look. The hippos were used to her presence on this side of the riverbank but not on the other, which meant we would have to be very careful not to get too close to them or to surprise them in any way.

Silas, Karen's assistant, tied a few logs together with rope in a semblance of a bridge. We laid it across two deep, fast flowing places in the river, balanced on rocks, and managed to jump on stones protruding above the running water until we were on the other side. This in itself was an adrenaline rush. I'd seen enough crocs in the last week on the Turgwe to know that they were virtually invisible in the water.

On the other side we scaled some smooth grey rocks on the bank, where long grass grew through the cracks and the water was shallow. Suddenly Karen stopped dead in her tracks. Without being told, I instinctively knew it was a snake. Karen had frozen solid and Silas and I copied instantaneously before backing away. I stayed about ten metres away while Karen went in for a courageous closer look. The snake had been absorbing the sun's rays on the large, flat rock in our path and was now slithering into a hole under some grass.

'Egyptian cobra,' Karen muttered, looking slightly unconvinced, 'but I didn't get a good enough look at him. It could've been a mamba. A good two metres long that one.'

It was the middle of the Zimbabwean winter, a time when snakes were meant to be hibernating, weren't they? I convinced myself that this was an insignificant aberration to the rule and that we were unlikely to walk into any more. It was all in a day's work for Karen. She was actually enjoying herself.

We continued our search for the hippos, silently stalking through the long grass. In some places the grass was almost as tall as I was. All of our senses – sight, hearing and smell – were on high alert. We couldn't afford to walk into the hippos here, not to mention any elephants or leopards that might be moving through the area. The riverine jungle enveloped us and a sense of not knowing what was around the next corner triggered an automatic response from my adrenal gland. With every step, the circumstances could change.

After about half an hour of stalking we stood on a large, elevated rock and looked through our binoculars for any movement. We listened intently for any telltale signs: the rustle of grass, the snort of a hippo, the cough of a leopard. The bush was far from silent. A warning sound that might mean the difference between life and death could be easily lost in the babbling of birds, the roar of the rapids and the whisper of the wind in the trees.

Karen pointed to something in the distance. I raised my binoculars and saw movement. The fuzzy demarcation of a hippo's head showed itself through a tiny shard of sunlight in the forest. They didn't know we were there and they were coming towards us very slowly, intent on their grazing. I was still looking in their direction, hoping for a clearer view, when I heard the crackle of grass moving. The noise was coming from in front of Karen, just a couple of metres away from her feet. Then I glimpsed a shiny flash of snakeskin slipping through the grass. Triangular shapes in shades of brown glistened as the reptile moved, slowly and sinisterly, through the undergrowth.

Karen didn't seem to have seen it, so I hissed, 'There's a snake in that grass,' as I backed away.

Unperturbed, Karen leaned over for a closer look.

'It's a puff adder,' she said simply, 'and it's big eh? It's actually very beautiful.'

By now I had thrown the theory that snakes hibernate in winter out the window. Nothing is *normal* in Africa. After that, anything that made any noise or rustled in the grass made me jump involuntarily. I was watching my every step, expecting a snake to be under every clump of grass. Karen and Silas were now more alert as well, and when a francolin blasted out of a bush in front of us at breakneck speed the small bird took several years off all of our lives simultaneously!

Such experiences are supposed to be character-building but by the end of it I wasn't having much fun, which means it must have been a good adventure.

One morning Karen and I jumped into her Peugeot to drive down to Bob's pod. She always parked her *bakkie* on a rise so that we didn't have to push it to get it going. The handbrake didn't work, so it was my job to pull out a brick from behind the back wheel so Karen could let it cough and splutter into action. It wasn't a four-wheel drive, which meant that at certain times of year, when the rivers and their tributaries were running, Karen was stuck at Hippo Haven.

We were spluttering along down the track, approaching a dry riverbed, when I spotted the unmistakable shape of a snake on the road.

'Puff adder!' I cried.

Karen pumped the brakes but we were going downhill and the brakes were only slowing us down, not stopping the car. She pumped her heart out and the Peugeot pulled up just a metre away from the snake. The puff adder hadn't moved at all, despite the looming vehicle. I shuffled into the driver's seat and kept my foot on the brake so Karen could coax the lethargic creature off the track. But the snake had other ideas.

Suddenly it had the energy of a marathon runner and it slid feistily into the car wheel on the driver's side. We were on such a steep slope that I didn't dare move my foot from the brake, lest the car run over both Karen and the snake. I tried not to think about the stories I'd been told of snakes that crawled up under the underbellies of cars and into the seats of the cabs. Karen's car, of course, had numerous holes in it, both inside and out, through which anything of a slithery nature could enter. I started to sweat. I was losing feeling in my toe from holding it so firmly on the brake. Karen was using a stick to try to pry the snake out, but now, to my dismay, she said she couldn't see it any more. It was starting to rain.

While all of this was going on, a Land Rover appeared. It was Alistair Pole, the local African wild dog researcher.

'I've come to pick Tammie up,' the strikingly blond man said, greeting Karen in a charming accent that seemed to be a mixture of Zimbabwean and Scottish, 'but it seems she's otherwise occupied.'

As if I didn't have enough on my hands with the puff adder and the car without brakes, now there was an unexpected gorgeous male researcher in the equation. With no radio, we'd had no warning of his arrival. It was all too much! I leaned out the window to greet him and tried to smile. Chuckling, Alistair joined in the hunt for the missing puff adder. I still wasn't convinced it wasn't about to squirm onto my lap from under the driver's seat. Alastair opened the bonnet of the car, but the snake was nowhere to be seen.

'You'll just have to drive and see how it goes,' he said.

Karen agreed. We might spend all day looking for the snake otherwise. My aching toe and fluttering heart supported the decision.

Karen shuffled back into the driver's seat and let the car run downhill a little before the engine lurched into life. Sadly, the unfortunate snake had, after all, climbed into the engine and experienced a fatal encounter with the fan blades.

We dashed back to Hippo Haven, where I flung my wet clothes from the washing line into a plastic bag and waved goodbye to Karen as I drove off with Alistair, the 'wild dog man', and his French volunteer, Nico, for my next adventure.

Each day with Alistair began in the dark pre-dawn when we headed off in his Land Rover to do what he called 'prey counts'. Part of Alistair's work, he explained as we drove along the wide, gravel roads of the conservancy, was to determine what species and how many of them were killed by wild dogs.

The African wild dog is an endangered species and continues to be persecuted throughout much of its range. Humani was one of twenty-two farms in the Save Valley Conservancy, a jointly managed area of over 3400 square kilometres. Although they were now game farmers, some cattle-farming instincts still ran deep and one of these was an inherent hatred of the wild dog. Some of the farmers in the conservancy still shot wild dogs, Alistair said, because they believed that they ate too many of their expensive antelopes. Antelope, like nyala and kudu, fetched handsome prices from trophy hunters. As some of the farmers saw it, each male antelope eaten by a pack of wild dogs was a potential trophy bull, which meant money trickling out of the bank.

Alistair was discovering that impala formed the vast majority of the wild dog's diet. Impala were the most common antelope in the conservancy and he had shown that the wild dogs' impact on the total population was very low. Wild dogs probably consumed less than three percent of the impala population in the conservancy, which was much lower than the impala population's yearly recruitment of lambs.

By counting the prey species in the conservancy, we were providing a baseline figure that allowed Alistair to estimate the relative off-take by the wild dogs of the farmers' antelopes. With only sixty-one wild dogs in the conservancy and less than five thousand in Africa in total, this work was crucial to ensure the conservation of this simultaneously loved and hated species. Alistair knew he would have to provide hard facts to convince tough farmers that keeping the wild dogs could be good business because tourists liked to see them too.

I absorbed all of this information like a sponge. To a twenty-year-old, aspiring conservationist, can you imagine anything more attractive than a good-looking, down-to-earth researcher with a passion for wildlife and a great sense of fun? Khaki fever hit me with fury. I was fascinated by more than the wild dogs. Although nothing more than a crush on my side, and a friendship on his, ever developed between us, the short time I spent with the 'wild dog man' certainly convinced me that wildlife research was the right field to be in when it came to meeting attractive, intelligent men in the bush!

To me, the day-to-day grinding research of game counts was bliss. As we drove along predetermined transects, we recorded the species we saw, the numbers and sexes, the distance from the car and angle from the road, the speedometer reading and habitat type. I was allocated the job of scribe while Alistair drove and his trackers, Ruben and Nico, sat on the roof looking for animals. A hard tap on the roof would signal that they had seen an animal, at which point Alistair would slam on brakes and we'd make a recording. This went on for hours and hours, day after day, kilometre after kilometre.

For five days and five nights we didn't bathe. It was the height of winter and every night I froze at temperatures of around five degrees Celsius while sleeping out in the open on top of the Land Rover roof. The Landie bounced over the bumps and swerved around rocks on the endless dirt roads. On a couple of occasions we had to get out to move some tree branches that had been dragged into the road by elephants. I was dirty, often freezing, and exhausted: I was having the time of my life.

Wildlife filled my days and permeated my dreams. A small herd of plains zebra in a blur of black and white stripes galloped across the dirt road, then shied. Dust swirled around their hooves as all seven zebra came to an abrupt, simultaneous halt, as if they'd orchestrated the performance just for us. A black-backed jackal shimmied down the road beside us in the soft peachy light of sunset, absorbed in a world of smells and sounds of his own. As we drove through dense woodland along a track where branches were broken all over the road, we came upon a herd of elephants. We must have given them a fright because they ran away in an ungainly fashion, their tails in the air.

The elephants in Save Valley Conservancy had been reintroduced from the nearby Gonarezhou National Park. In Gonarezhou, which means Place of Elephants, they'd been subjected to unrelenting poaching and they still carried the memories of these negative associations with humans. It would take a few generations for the elephants to feel that they were safe from people. Elephants have long memories. Those in Save Valley were flighty and readily charged people in vehicles and on foot. But life prospered here, it seemed, even if the elephants were still wary. Bushbuck, giraffe, warthog, black wildebeest, bushpig and duiker — if it was possible to see it in Save Valley, we did. We didn't see any wild dogs, but that didn't matter in the slightest.

Alistair kept me entertained with his incredible knowledge of the bush and his unfailing sense of humour. He had been born in Zimbabwe, but schooled in the UK. He was doing his PhD through the University of Aberdeen after the Africa bug bit him and called him home. He had a strong work ethic and a deep passion for conservation, but as we downed a bottle of sweet Amarula liqueur he reminded me that doing research shouldn't be all work.

'You've still got to have a life,' Alistair said, with a serious frown.

Then, with a charming smile, he added, 'It's hard sometimes when you live out in the bush.'

I laughed, thinking how much I longed for this lifestyle. After two years of living in Brisbane, I knew that it was possible to be lonelier in an ant nest of people in the city than with a couple of people in the middle of the African bush.

'You know, it's really annoying,' I announced, 'I can't properly explain why I love it here. People back home think I'm mad.'

'Aahh, yes,' Alistair said, scratching his blond stubble. 'Africa-itis… It should be a recognised medical condition, you know.'

I couldn't have agreed more as I lay on a foam mattress on the roof of the Land Rover, gazing at the burgeoning night sky, aglow with the sparkling lanterns of the Milky Way. I was at home in Africa in a different way to Australia. If this was what being a zoologist was all about, then there was no doubt in my mind that this was what I wanted to do. I dreamt about it, day and night.

Before dawn on my final morning of game counts with Alistair I was awoken by the sound of gunshots. The reports rang out so loudly that I could have sworn someone was shooting at us. I sat bolt upright, then lay down just as rapidly, realising that if we were being shot at I'd be a harder target if I was lying flat on the roof of the car. Could it be poachers? I immediately thought.

Suddenly about thirty men in khaki uniform charged out of the bush from all angles and began jogging in military fashion towards the car. I blinked, wondering if I was still dreaming. The men were yelling in Shona in deep, commanding voices. Were we being ambushed? Were they going to take us hostage? What the hell was going on?

The men were still coming towards us like a throng of matabele ants on full rampage, their faces angry and determined, chanting like warriors. I peeked out from under the covers, hoping they wouldn't see me. I knew that Alistair and Ruben were on bedrolls on the ground beside the car.

Then, as they reached the car, the men continued their rampage straight past us, as if we weren't even there! I watched them go by with my heart in my mouth and my eyes big as golf balls.

I leaned over the side of the vehicle to see Alistair grinning while making a cup of tea over the gas cooker.

'Alistair? What's going on?' I asked, the terror in my voice surely evident.

'Scouts training,' he replied with an amused smirk. 'Anti-poaching guys.'

I climbed off the top of the Land Rover, ogled by three rows of men being given orders by a man in blue and white striped pyjamas. The men were standing about fifty metres away, lined up with their chests puffed out and their feet firmly together. Once I'd recovered from the initial shock, it was strangely comical. Here was the leader of the conservancy's anti-poaching brigade giving orders at dawn and looking like a Banana in Pyjamas! After he'd been yelling orders at them for a while, they ran off into the bush and up a nearby *kopje*. This was training, Alistair said.

When I nipped off into the bush nearby to the African ladies room, I noticed that all around us were bags of kit and sleeping gear. All night, to our complete ignorance, we'd been surrounded by the anti-poaching team sleeping soundly around us. As a result of our late arrival, after the anti-poaching team were already asleep, no one had warned us about the wake-up call we received in the form of gunshots.

This trip to Zimbabwe was a wake-up call in many ways. I knew now that I never wanted to leave. I had to make a future for myself there. I was going to become a wildlife researcher.

AN IMPALA INITIATION

I was in agony. Bloated and ruptured blisters and cuts created an angry mosaic of inflamed colours on both of my feet. Putting on my boots was pure, unadulterated torture. My nose was peeling. My sunburned face glowed a shiny pink like young beetroot. The muscles in my calves throbbed whenever I stood up. On top of this, I had a tracker who didn't appear to speak more than a few words of English, a vehicle without an operational battery and every one of my study animals ran a mile when they heard our rackety vehicle coming. And this was life as a wildlife researcher in Africa? This was what I'd been dreaming about for so long?

At the age of twenty-one and in my final year as an undergraduate, I had spent many months planning this return to Humani. I'd put a lot of thought into finding a rational reason to go back to Zimbabwe which would appease all the Aussie patriots who thought I was mad in my longing for Africa. My mother's friends were convinced I had a boyfriend there that I wasn't telling anyone about. Conducting the fieldwork for my Honours thesis in zoology seemed like a very rational reason to me. Well anyway, it was a damned good excuse.

In some ways, kangaroos and impala weren't so different from each other. Besides which, there were important management issues affecting this abundant ungulate in Africa that also applied to macropods in Australia. I was interested in one of these issues in particular – the effect of trophy hunting on the behaviour of animals. The controversial idea of shooting animals to save them fascinated me. I had no desire to be a hunter, but I knew from first-hand experience that farms like Humani, which operated trophy hunting safaris, had flourishing populations of animals. To business-minded farmers, giving animals a monetary value made them worth conserving. By conserving the habitat for trophy-hunted species, Roger was effectively protecting an entire ecosystem, a web of life that included trees, insects, birds and reptiles. I fully supported this form of hunting because it represented sustainable use of the natural resources. But I was curious to know whether the animals on farms that conducted trophy hunting and culling for meat differed in their behaviour from those on farms that only allowed photographic tourism. Were animals more skittish when they were hunted?

When I emailed Roger to ask his opinion of a study to look at the impact of hunting on impala vigilance behaviour, he told me he would sponsor my study with a free vehicle and tracker for the duration of the project. Armed with nothing more than a crazy idea, a sponsor in Zimbabwe and a lot of inspiration, I convinced the university to let me investigate the issue for my thesis and asked zoology professor Anne Goldizen to be my supervisor.

I'd been inspired by Anne's lectures during my undergraduate studies. An American who had adopted Australia as her home, she had done her doctorate on monkeys in the remote jungles of South America a generation after one of my heroes, Jane Goodall, had begun her pioneering research on the chimps of Gombe in Africa. In my eyes, Anne was a modern-day Jane Goodall, with a kind, open face and a sense of adventure.

Some people believe that when you want something enough, and if it's what you're meant to be doing with your life, the universe and everything in it conspires to help you achieve your dream. This was certainly happening to me.

Once again, during my summer holidays, I rose before dawn each day to pick asparagus. It was backbreaking work that required mental stamina more than physical strength, which is why, my boss Sok Kienzel told me, he employed only women for the picking. With every asparagus I tossed into my red plastic crate, I visualised the dollars accumulating in my bank account. During the university year I'd packed shelves at a local convenience store a couple of nights a week, riding my bike there and back, up and down the hills of St Lucia. I'd worked damned hard to earn the cash to pay for my flight and the university had contributed enough to pay for my fuel in Zimbabwe. Then the South Pacific chapter of Safari Club International, a worldwide hunting organisation, chipped in with some money too. It was all happening!

I was armed with a heavy pile of scientific papers on previous studies of vigilance behaviour on every creature under the sun, from ostriches and ibises to mountain goats and springboks. Vigilance behaviour is simply when an animal scans its environment, usually looking for predators, but sometimes for rival males or potential mates. This is a pretty important activity for prey animals because they need to be constantly on the alert for threats to their lives. In a place where things can eat you at any moment, you're either vigilant or you die. In my study I wanted to see whether being hunted by human predators made impala more vigilant than those subject only to natural predators.

AN IMPALA INITIATION

I must admit that impalas didn't excite me much. A medium-sized antelope, impala have a rufous-coloured coat and lighter coloured underbelly. The males have lyre-shaped horns, which they use to defend themselves from predators and to fight with other males during the rut or mating season. They are elegant and sprightly animals. The reason they didn't excite me was because they are extraordinarily, boringly *common*, hence the name 'common impala'. Found throughout southern, eastern and central Africa, they have adapted to a wide range of environments, and this ecological flexibility is probably what accounts for their success as a species. Predators love impala. Everything eats them, including humans, and locals don't bat an eyelid when they see a herd in the bush, unless they feel like an impala steak. There's a reason why Zimbabweans call impala Zambezi goats.

Like all first-time researchers in Africa, my eyes were blinkered by grand and unrealistic aspirations of working on one of the Big Five. Initially I was tantalised by lions and leopards, fascinated by elephants and awed by buffalos and rhinos. But to be honest, when it came down to it, I didn't care what species I worked on as long as it was in Africa.

It made good sense to work on a common species, especially as this was my first attempt at solo research in the field. But where was the fun in that? I wanted an adventure with all the dangers and excitements of working on an African mammal with grunt. I shouldn't have worried. What I didn't realise was that when you work on a prey animal, predators always surround you. Just being in the bush looking for impala, I could encounter anything from hyaenas taking down an impala lamb to elephants plodding down the road. I would have to learn to be very vigilant for predators myself.

Back then I thought my budding relationship with impalas was just going to be a fling before the real thing came along. I could never have imagined that this species would be the focus of the next five years of my life, that impalas would become The One.

On my arrival at Humani in late November 1998, the beginning of the wet season, Sarah told me that I would be staying with her at her house. She had moved out of her parents' house after deciding that working for and living with your father wasn't such a good idea after a while. She explained that my timing was impeccable because she was in great need of someone to drink wine and gossip with. I was honoured. Our days as Scary and the Aussie Slave were in the past. Having survived that era, we had become close friends.

Sarah's house, like her parents', was a colonial-style oasis enclosed by a tall, elephant-proof wire fence. The sprawling green grass and tall trees around the quaint cottage gave the place an English feel. The white lattice outdoor table and chairs set under the thick canopy looked as if it was made especially for two chicks to drink wine at. And the entire house, which we christened Scarydom Castle, was guarded not by sharp-fanged dogs but by four terrifying white geese. As I carted my bags inside, they chased us with their menacing beaks, while squawking murderously. Sarah picked up one of my bags and ran like billyo. On cue, I did too, running like the clappers at her heels like a trusting younger sister. She was the youngest of three daughters in her family and I was the eldest sibling in mine, but we had swapped roles with ease.

'Scary, why on earth are we running away from your geese? Aren't they meant to be keeping other people out, not keeping you in?' I yelled as we ran inside, the geese chasing after us and hooting malevolently.

'Oh look, just shut up, you horrible battleaxe, and let's have a drink of wine!' she rebuffed as she made it to the door.

Later on, Roger introduced me to my research team: a tall Shona man by the name of Ipheas, and the Blue Beast, a brakeless,

AN IMPALA INITIATION

roofless, doorless, batteryless, suspensionless jeep. Ipheas looked to be in his forties and on first impression came across as fairly serious. He couldn't have been too impressed by the idea of working with a madam *mukiwa*, a white woman, spending his days looking for Zambezi goats. He must have been expecting to be bored to tears. Besides that, he didn't seem to speak much English besides 'hello', which meant that we would have to get by on my meagre command of Jalapalapa and some sign language.

I jumped into the driver's seat of the Blue Beast and a rusted, coiled spring vaulted out of the torn, stained upholstery. I put her into second gear and held my foot on the clutch while Ipheas gave her a mild push from behind. When she had a bit of speed on the downhill run I let go of the clutch with a screech and pressed my foot onto the accelerator. She coughed. She spluttered. She farted clouds of billowing black smoke. Then she rumbled into action as if it was the most awful thing she'd ever been compelled to do. As we drove off, she smoked furiously and choked often before the diesel engine spluttered cantankerously back to life.

My first thought was how the hell was I going to see any impala with the racket this car made? Apart from the roaring engine, anything that had the capacity to shake was rattling and creaking and screeching with a passion. Any impala with a fair sense of hearing would hear us coming a mile away.

Undeterred, a solemn-faced Ipheas and I set out on our search for impala. He was sussing me out as much as I was him. I crunched the gears up and accelerated until we were soaring at a top speed of about forty kilometres an hour. Without any roof, doors or windscreen, the wind blew my hair back and the sun smiled warmly on my face. It was good to be back in Africa, I mused, glorious in fact! I revelled in the organic smells of the bush, the freshness in the air after rain overnight. I had a tracker, a real, live one called Ipheas. I had my car, a true bush car with character. And I was doing research in Africa. *Yeah*!

Caught up in my reverie, I failed to notice an enormous puddle covering the entire road ahead of us until I was almost on top of it. I pumped the brakes desperately and realised that they didn't like to be pushed into anything quickly. My eyes widened in horrified anticipation.

'Oh madam, oh madam!' Ipheas exclaimed, shaking his head, not daring to look away.

I shut my eyes and held tightly onto the steering wheel as we inevitably hit the pool of muddy water at full speed. A massive splash of brown water engulfed the entire vehicle, and we didn't have a roof! Ipheas and I looked at each other, our faces dripping with muddy water, and burst out laughing.

Well, I may not have earned my tracker's respect on our first day of working together, but it looked like we were going to have some fun. The ice was broken.

After a couple of days of searching for impala at Humani, I realised that my carefully constructed plans would need to be changed. We weren't having trouble finding the impala. They were as common as flies in the Aussie outback. The problem was getting close enough to them for long enough to record their behaviour during a five minute sample. In theory five minutes isn't long; in practice it wasn't so easy.

The Blue Beast made such a racket that when we spotted an impala herd they were already running away from us. We wanted to record natural behaviour when they were relaxed and foraging, not running away in fright. At Senuko, the photographic tourism property neighbouring Humani which I was using as a comparison, we had the same problem. Impala were abundant, but hell, they weren't stupid!

AN IMPALA INITIATION

In my arrogance I hadn't even considered the fact that an animal that is hunted, stalked and killed by most predators in the African bush might be a little bit smarter than I was. I'd totally underestimated them. We were in their environment, where they had the advantage. They were used to being hunted by the master predators like lions and leopards. An Aussie girl in a rackety jeep was a laughable threat. I imagined them chuckling and joking with each other as they pranced away in elegant bounds time and time again: 'Stupid girl, she thinks she's got us sussed!' My respect for impalas began to grow.

I was also learning an important lesson about stalling the Beast: don't. As we drove home one afternoon, still without a single viable behavioural sample, a herd of zebra galloped across the road and into the savanna. Delighted, I pumped the brakes and pulled out my binoculars for a closer look. By accident, I dropped the clutch and the Blue Beast did the only thing that she ever did fast: she stalled.

We were at the bottom of a gully on a steep road that rose up on both sides of us. With great humility and patience, without saying a word, Ipheas got out of the car and started to push it uphill. It was close to forty degrees and sweat drizzled down his brow. I jumped out and helped him. I felt like an idiot. Together we shoved the Beast uphill, our arms and backs straining with the effort. Then, when Ipheas was satisfied that we were high enough, he held her there while I put her into reverse. She rolled backwards with gusto, gathering momentum as she went. I let go of the clutch while trying to manage the temperamental steering wheel. To our mutual amazement, the Blue Beast started and the sound of her motor droning into action was like music to our ears.

As for the impalas, I was determined not to let them beat me. Although most studies of vigilance behaviour on African ungulates had been conducted from vehicles, this only worked where animals were accustomed to the presence of humans and cars. At Humani and Senuko, the impalas weren't habituated as they often are in national parks with lots of tourist traffic. If I was going to have any success, I would have to play the game their way. I would have to think like a predator. Ironically – or perhaps not – I would have to become a hunter.

Stalking a herd of impalas on foot was a complex process that required complete concentration. We would park the car a kilometre away from where we suspected we'd find a herd, and then, importantly, check the wind direction. We stalked quietly, Ipheas in front, me behind treading in his footsteps, trying to stay downwind, being careful not to make any sounds by crunching leaves or cracking sticks underfoot. A crackle or a crunch was all it took to give us away. Only one impala had to spot us or smell us and emit an alarm bark for the whole herd to head for the hills in the time it took you to swear (which I did – often).

After scaring away several herds, I started to get the hang of stalking and Ipheas started to understand what I was trying to do. It was hard to explain to him in haphazard Jalapalapa that I wanted to understand the impalas' behaviour so that we could learn more about the effects of hunting. Like most Zimbabweans, hunting was like breathing to him. It was something you did to survive. He'd never heard of anything like an animal rights group. What's more, the luxury of spending several months watching impala for the reward of a piece of paper from a far-off university must have seemed crazy. In fact, I'm quite sure Ipheas thought I was off my rocker.

If we successfully stalked a herd, we sat quietly behind a bush and chose a focal animal, recording its behaviour from about fifty metres away. I started the five-minute timer and watched the animal intently through my binoculars. With my other hand I clicked another timer on and off as the animal raised its head in between feeding bouts. Each time it raised its head, I said 'Yup'

AN IMPALA INITIATION

to Ipheas, who clicked a counter that measured the number of times the animal looked up. This gave us two measures of vigilance.

Sometimes our focal animal would get lost in the herd or walk behind some bushes, so we'd have to cancel the recording. It was slow and tedious work that required a lot of concentration and a little bit of luck. There was also an element of danger, given that we were watching a herd of prey animals and predators could be watching them, and us too.

After a few weeks we were really getting the hang of stalking impala; however, just when you think you've got everything under control, Africa strikes again. It was the height of the wet season in December. With each day it seemed to rain more and for longer. On most days, rain pelted down in late-afternoon showers that carried on into the night. The deafening thunder shook my soul with its ferocity. Sometimes it rained all day.

The roads turned to rivers and then to mush as the water seeped into the earth. Just keeping the Blue Beast on the tracks was a mission in itself. My arms grew muscles in places I never knew existed. I battled to control the jeep as she slipped and slid and sunk. I needed to keep a little bit of speed up to get through the muddiest patches, but this made her even harder to manage with a steering wheel that had a mind of its own. Skidding even a centimetre off the main tracks resulted in the whole vehicle sinking into the mud up to its axles. The four-by-four training Dad had given me during family camping trips hadn't prepared me for this.

The first time we got properly bogged was one of the hottest days I can remember. All four wheels were stuck in the tracks in mud up to the axles.

'I make a plan!' Ipheas announced with grim determination each time we tried and failed to get her out of the mud. He said that about four times.

Both of us were covered in the brown sloppy goo. It was in my hair and deep under my nails and up my nose. Our radio was dead so we couldn't call for a tractor to tow us out. We had to give it our best shot.

But after trying for two hours, flinging mud out with our hands and poking sticks under the wheels to give her some traction, Ipheas looked up, grinning, and said resignedly, 'Aahh...today, I killed a lizard!'

His face was beaded with sweat. He was smiling with good humour, despite his exhausting efforts. I raised my eyebrows and smiled back, amazed at his tenacity.

'It is just an African joke,' he laughed, which I took to mean that he'd worked bloody hard for not much reward!

After three hours we resorted to walking back to get help. It was midday and the humidity was choking. It was close to a ten kilometre walk to the croplands where we knew there would be a tractor to pull us out or a radio to call for a tow. There was no other choice.

We started walking through the heat haze. I carried only the water bottle in my hand and a wide-brimmed cane hat on my head. Ipheas was wearing heavy canvas overalls as all the men did, no matter how hot it was. He must have been sweltering, but he was acting as though he did this every day of the week.

As we walked, I fell into a kind of trance. Perhaps it was a heat-induced stupor. The water in my water bottle grew warm. It tasted sweeter the warmer it became, but quenched my thirst even less. I couldn't have consumed enough water to match the amount of liquid I was purging through sweat. I recall the deafening whirr of cicadas in my ears as I plodded behind Ipheas. Dazed by the sun, I stared at his mud-caked black feet with their pale soles, taking step after step, rhythmical and hypnotising, making large indents in the sand. A combination of sunscreen and sweat dripped

AN IMPALA INITIATION

into my eyes, stinging them acutely and blurring my vision even more. The blisters on my feet soon forced me to take off my boots and walk barefoot as Ipheas was. The thorns on the track were preferable to the pain induced by my swelling feet. An hour and a half later, we reached the crops and radioed Roger for a tow. I remember thinking, I'm glad that's over and I'll bet that won't happen again for a while. Boy, did I have a lot to learn – we were about to start getting bogged every single day!

No matter what happened, Ipheas always found something to laugh about. He was a gentle and humble teacher. One time we spotted the silhouettes of four big cats in the distance. At first I thought they were lions, but Ipheas, with his formidable eyesight, laughed at me. Through my binoculars I could see clearly that they were cheetah.

Amazingly, the very next day we saw cheetah again, this time right beside the main road.

'You are good luck,' Ipheas said to me, grinning widely.

I didn't slow down at all as we drove past. In fact, I sped up because I thought a wild cheetah could be dangerous and this one was less than five metres from the vehicle.

Ipheas laughed again and said, 'Aahh... *Indingwe* (cheetah)... They are no problem.'

I felt a bit stupid for accelerating past a non-threatening but extraordinary sight, but Ipheas didn't rub it in.

On another occasion I made the mistake of not checking the fuel level and we ran out on the way back to Humani from Senuko. The fuel gauge didn't work and the indicator was permanently stuck on empty, but in any case I should have filled up. Ipheas always carried the mobile radio on him so I asked him to radio Humani to ask someone to bring us some fuel.

He shrugged his shoulders and replied, 'No, madam. Radio *kufa*.'

The radio always seemed to die whenever we needed it. I swore.

'*Ndino famba.* I walk,' Ipheas announced decisively, getting out of the car.

Before I had a chance to object, he was already running along the road back to Humani. I couldn't believe it. It was at least a twenty kilometre run back to the house. It was close to the middle of the day and boiling hot. Ipheas hadn't even hesitated!

As I waited beside the road under the shade of a tree I realised Ipheas may have deliberately set off before I could protest. I would have only slowed him down. Poor Ipheas was paying the price for my mistake. I waited…and waited…and waited…

After about an hour I heard the distinct rumble of a car's engine. At first I thought I was hallucinating. But there it was, a shiny Pajero coming towards me and heading in the direction I needed to go. I leapt out of the bush and ran out onto the road like a mad woman. Red-faced under my cane hat and probably babbling, I explained my predicament. The car full of South African holiday-makers were only too happy to give me a lift. I was so relieved I could have kissed them! They made some room for me on the back seat in the divine, airconditioned cab where one of them started filming me on their home video camera and barraged me with questions. I was too grateful to protest.

About ten kilometres down the road we found Ipheas still jogging like a marathon runner. He grinned from ear to ear when he saw the car. The South Africans passed him a freezing cold Coke and I told him to wait in the shade while I went to get a jerry can of fuel. Ipheas waved us off looking like all his Christmases had come at once.

I was aiming to record a minimum of five vigilance samples a day. This had seemed an impossible task when I'd arrived in November. Initially I'd been ecstatic just to record two samples. Now, in January, I was recording up to nineteen in a day. Ipheas

AN IMPALA INITIATION

and I were working together as if we'd done this all our lives and we'd seen wildebeest, giraffe, eland, kudu and warthog at close range, the experiences made all the more magical when the animals didn't know we were there. Then we could watch them as they carried on with their natural behaviour.

One day Ipheas and I were watching a herd of about a hundred impala grazing on the short, green grass of Eland Pan at Senuko when two black-backed jackals arrived on the scene. They trotted up to about ten metres from where we sat in the grass, not moving a muscle. With all the baby impala running around, the jackals would have been having a field day. When the female impala lambed in the wet season, all within a couple of weeks of each other, it was like fast food – African style.

This particular pair of jackals had no idea we were watching them. Their short black, tan and white coats made them virtually invisible in the bush, but we were so close that we could distinguish individual hairs on their faces. They both lay down, preparing for a snooze. Then the male lay on his back and began to lick and groom his white furry underbelly. He rolled around, trying to scratch a spot that was just out of reach. I turned my head slowly towards Ipheas and smiled. As usual, he was already grinning. It was a magical moment.

Of course, jackals aren't generally a threat to humans. When it came to our safety, Ipheas called the shots and he didn't take any chances.

Late one afternoon we were walking around Eland Pan at Senuko when Ipheas said, 'I think we go now because of *shumba*.'

He seemed a little edgy. That morning we'd seen fresh lion spoor in the same area, huge, deep paw marks in the mud. I didn't question his instincts and we started heading back to the car. Apart from the spoor of predators, such as footprints and dung, evidence of their ubiquitous presence was often evident when we found the scattered bones or decaying carcasses of prey animals. On one occasion we picked up a dead impala lamb. Jackals had chewed off its leg and vultures had chewed the face to the bone, Ipheas explained in sign language.

As we drove off on this day, Ipheas pointed to a spot about two hundred metres from us, on a rise. There were about six cheetahs, sitting up and watching us with interest. Their spots weren't visible at that distance, but their lithe, sinuous shapes were now obvious to me. I wasn't threatened by them. It was just incredible to see. But it was a timely reminder that when the sun is setting and you are working in an area with a concentration of prey animals like impala, you are entering the realm of the predators.

As we drove back, the grey sky began to sprinkle gentle raindrops upon us. We slipped on our raincoats and drove on. We were used to getting rained on while driving home. It was all part of a day's work. We could always put up the cracked glass windscreen to mitigate the impact of the rain if it got too heavy to see where I was driving. Ipheas almost always saw the rain coming before I did. With rain and predators, he seemed to have a sixth sense for their imminent arrival.

He would look up to the sky, an ominous expression on his face, and then announce with a smile, '*Mvura uya*. Water/rain comes.'

I would ask him, '*Manje?* Lots?'

Usually he would respond with a knowing smile, 'Ah-ha.'

On this day, however, the rain was nothing more than a drizzle. I was focused on the road ahead and thinking of how good a glass of dry red would taste when I got home when the fuzzy shape of a lone, small animal became visible in the distance. It was standing in the middle of the road and it looked like a jackal or a dog of some kind. I blinked raindrops out of my eyes and opened them to see my first sighting of an African wild dog. Gobsmacked, I slowed down and let the rain coat my face with

AN IMPALA INITIATION

moisture as I watched him. A solitary male, he pranced around on the road, watching us with inquisitive eyes. His shaggy patchwork coat was dappled and glittering with drizzle. I now understood why people called wild dogs 'painted dogs'. Then the he gambolled away in a happy-go-lucky fashion as if he had better things to do with his time.

As well as the rain, the wet season brought with it a proliferation of insects. The rocketing humidity provided the perfect environment for their breeding. Anne, once a nurse, instructed me to cover my legs in vaseline after they came up in hundreds of tiny red bites. The bites weren't itchy, but I looked as though I had contracted some terrible disease.

For some bizarre reason, all of the cats and dogs started licking my vaselined legs as though they'd never tasted anything so delicious. No matter how many times I shooed them away, they continued to slobber on me with rough, sloppy tongues.

'Can you blame them?' one of the Humani bachelors said. 'They've got good taste.'

I was covered in animal saliva and didn't appreciate the joke. I'd become used to the fact that I was a young woman doing what was considered a man's job. I was aware that I was walking further and working much longer hours in physically tougher conditions than many of the Humani bachelors and I think they knew it too. None of the men hunted at this time of year: it was too hot and too wet. Perhaps I was a curiosity to them, being so different from the women they knew who took on more traditional roles. In their eyes, I was a feminist and maybe even a lesbian. From the attention I received from a couple of them, it wasn't exactly a turn-off. But the truth was, I didn't want to pour tea, bake cakes and type letters. I didn't want to conform to traditional expectations just because I was a woman. I had had the benefit of an open-minded upbringing and a good education which had taught me that I could do anything I wanted.

After the bites on my legs, a strange white pustule appeared on one of my toes. It was constantly sore and throbbing. Sarah told me that a putsi fly had probably laid an egg in my foot and I would just have to let it grow until the larvae came out. The thought of a maggot growing inside the pussy blister on my toe made me feel physically ill.

Ipheas and I had been shedding our boots each day to wade through the Njerezi River, which the Beast could no longer cross, to our study site at Senuko. Rushing water flowed past up to my knees. I'd been walking in mud, sand and grass in our quest to get to the impala viewing areas. At one point it didn't stop pouring down for seven days straight. If we'd let the rain get in the way of our data collection, we would never have left the house. So I put up with the putsi fly larvae growing fat in my foot and the zillions of pink bites on my legs. It was a small price to pay.

With each day, we came up with more novel ways of getting ourselves out of challenging situations. I was learning first hand the African rule: make a plan.

Whenever Ipheas announced, '*Lomotor inashoopa*. The car's got a problem,' he may as well have said 'get ready to walk'. One time, when our radio, which served more of an aesthetic than a functional role, wasn't working again, we were walking back to the croplands along the main road when a Shona man came along on an old fashioned bicycle.

'Is that your bicycle?' I asked him, hoping he spoke English.

'Yes, it is,' the man answered.

'Please can I borrow it to go to Humani?'

He replied, 'Ja, okay. But you must bring it back on the vehicle.'

I thanked him profusely for his generosity and left the two men standing there talking in Shona.

AN IMPALA INITIATION

'*Handei*! Go!' Ipheas exclaimed and clapped as I rode off, chuckling unselfconsciously.

I was barely able to touch the pedals with my vertically challenged legs as the rackety bike's thin rubber wheels powered through the thick sand. The wonky contraption was wobbling precariously and I was puffing heavily with the effort. After about three kilometres of riding, I was getting into the swing of it. Then the chain fell off. I jiggled and joggled, trying to get the infuriating metal chain to go back into place. I swore. No one was around. I swore louder. Defeated, I left the bike on the side of the road and started walking. Foolishly, I'd forgotten to bring my water bottle. The sun struck the back of my neck with ferocity. Some days the temperatures had risen up to the forty-degree mark, but it was the humidity that was the killer. It created a sort of claustrophobic psychosis as it hung all around you, threatening to condense your dripping body into liquid skin.

To my relief, a few kilometres later, I noticed one of the game-scouts coming towards me on a motorbike. He offered me a lift and I willingly jumped on the back. As we headed towards the house, another vehicle approached. It was Chimeni, another one of the game-scouts, coming to our rescue. He'd heard our radio calls for help, but we hadn't been able to hear his response on our defective radio. I gave Chimeni instructions on where to find the bike and he said he would fix it and take it back to the man who'd kindly lent it to me, sending along my apologies and thanks.

There are many, many ways to get stuck in the bush. Ipheas and I discovered all of them. Our days were filled with adventures and sometimes my nights were too.

One evening, while Sarah was away in Harare, I lay in bed reading, alone in her house. A woman's cry rang out in the silence of night. She was wailing at the top of her lungs, screaming as if she'd been bitten by a snake or a child had been killed.

I ran out the back of the house to where Tongai lived with his wife and small children. The woman was wailing at the top of her lungs, a desperate cry for help.

I called out to Tongai, 'Is everything all right, Tongai? Is someone hurt?'

I was unnerved, to say the least. Tongai ran up to the back gate, looking a little unsettled too. I shone the torch around him, trying to work out what was going on. There was no sign of the woman, who I assumed was his wife.

'Ahh…no madam,' he said, sounding as though he was trying to placate me. 'It is just our culture.'

My immediate thought was that the woman was in a trance of some kind. Perhaps it had something to do with black magic. Maybe that was way Tongai was behaving suspiciously.

'She is in the spirit world,' Tongai continued, calling out over the top of her screams.

'Okay,' I said. 'She is not hurt then. Okay… How long does this go on for?'

'Oh…' Tongai thought for a moment. 'Maybe a year.'

A year?

'No, I mean tonight, Tongai? One hour?'

'Ahh no, maybe twenty minutes…half an hour…' he went on, then changed his tune and said, 'madam, can I have the spotlight? I will take her to my parents.'

I tried to smile, but I was shaking. Something about this didn't feel right. Tongai was carrying a little boy on his hip and he looked nervous. He went back to his hut and I could hear him talking to his wife, apparently trying to calm her down. Finally she stopped screaming and began sobbing. Her plaintive crying was terrible. A feeling of being very alone and vulnerable engulfed me. My heart was beating at a hundred miles an hour. The woman began taking huge breaths and blowing it all out in long, mournful sighs. When

AN IMPALA INITIATION

Tongai spoke it sounded as though he was reprimanding her rather than consoling her. She continued to moan. Then the night became silent again. Deathly still. Hours later, I fell into a light sleep.

The next morning I asked Ipheas about the incident. He told me outright that Tongai was lying. He punched the air and then punched the palm of his hand to show me what he thought had been really happening.

'He is beating his wife,' Ipheas said, shaking his head.

Sadly, wife beating is not uncommon in Zimbabwe. Women, especially those in rural areas where education is poor, often bear the brunt of their husbands' fists. It is not an easy life for an African woman. Years later, after I had been caught by the police for not completely stopping at a stop sign, the black policeman told my boyfriend at the time, 'You must beat your wife.'

I liked Tongai, but the thought that he may have been abusing his wife made me feel nauseous.

***

Towards the end of my study I joined Roger on a drive up to see the flooding Turgwe River, where the water had risen so high that no one could pass through on the main road to town. When Roger stopped at the river he noticed a movement in the bush near the car. He got out without saying anything and walked into the bush. A minute later he returned. In his arms he carried a very young kudu calf. Like Bambi's, the small kudu's deep brown eyes were a picture of endearment. The tenderness in Roger's eyes spoke a million words: the little antelope had clearly touched the heart of the lion.

He put her in the back of the Cruiser where I held her legs and covered her eyes with a cloth. It was only a ten-minute drive back to the house. Anne had saved baby animals before, maybe she could save this one. I felt hope surge through me. But sadly, as I held the little kudu's legs, I felt them stiffen. And I knew, even before I removed the cloth to see her glazed eyes, that she was no longer alive.

Once again, I felt my own departure from Humani looming. But there was hope because this time I knew I was coming back. I had enough data to write my thesis, which was a relief, but I also had something much more important. I now knew that I had what it took to do independent research.

I'd been bogged close to a hundred times, walked hundreds of kilometres, been towed out by a tractor a dozen times, ridden a racing bike through sand, crossed rivers barefoot, learned to stalk impala, been rained on so hard that I couldn't see where I was driving and even had my foot run over by a car (luckily it wasn't broken). I had thought I wasn't smart enough to do a PhD, but my Honours project had been a recipe for learning and now that I'd had a taste of it, I wanted more. What I realised was that doing a doctorate has more to do with having determination and drive than it does with being smart. If I wanted to make a difference in the field of wildlife conservation, as a woman and as a foreigner, I knew I had to get a decent qualification under my belt. And that would mean working my way up from the bottom.

SERENDIPITY

Serendipity is when happy things arise accidentally. That is how I came to Africa in the first place – by accident – but it's also how I came to leave Zimbabwe and move to Namibia. Most people I know have never heard of Namibia. Most Aussies call it 'Nanimbia' or think it's Libya when I tell them where I live. Fair enough. It's certainly a country that's miles off the beaten track for most travellers. Namibia is one of southern Africa's best kept secrets: a land of mystery and magic. The driest country south of the Sahara, Namibia is a desert land. I loved Zimbabwe, with its riverine jungles, wide running rivers and smiling Shona people. I had no desire or intention to venture anywhere else. Zimbabwe felt like home to me and Humani was a playground for learning. The Whittalls were like my surrogate family. But life had other plans for me.

Preparations for my PhD at Humani were going so well that I began to wonder whether it was all too good to be true. With the first class Honours I had received for my impala vigilance project, I had been offered an Australian Postgraduate Scholarship for my doctorate. Several knowledgeable people had tried to convince me to carry on with impala research for my PhD, and although it wasn't exactly what I'd imagined, I was slowly coming around to the idea. After all, it really didn't matter what I worked on. As long as the animal was in Africa, it would be an adventure.

My decision was made when, by extraordinary coincidence, I met Professor Peter Jarman at an animal-behaviour conference in Armidale. A grey-bearded man in casual pants and a red Kathmandu jacket, Peter was warm and friendly. He had a trans-Atlantic accent, a mixture of African, English and Australian it seemed, and twinkling eyes, and as we spoke about Africa I felt as though I'd met a kindred spirit. Until then Peter had simply been a famous name on many of the scientific papers I'd had to read for my Honours project. He was – and still is – one of the world experts on impala ecology. Although he had spent a lot of his life in eastern and southern Africa studying impala, Peter had made his home in Armidale and was teaching at the University of New England. He seemed as surprised to meet an Australian student interested in impalas as I was to meet an impala guru in Australia. The synchronicity was perfect: I would be one of Peter's last students before he retired from academia.

One thing that really did matter to me was that my doctorate would have some realistic, practical application for managers in Africa. There are so many pressing issues facing the conservation of wildlife and biodiversity, so many species becoming extinct through human influence and, often, so little communication between academic researchers and managers in the field, that I didn't want to produce a thesis that had no value other than a purely academic one. My doctorate would take me three to four years. I was determined to do something worthwhile in that time.

Peter, Anne Goldizen and I designed a project that investigated the habitat preferences of common impala in the Save Valley

Conservancy. Preservation of the right habitat is crucial for any animal's conservation, so I saw some value in this approach. Common impalas weren't endangered by any stretch of the imagination, but they were an important species for meat. Scientific thinking didn't come easily to me; I had to train my mind to become more analytical and critical, to pay more attention to detail. I wasn't a natural scientist and I'm still not. I was leading with my head where the science was concerned, although it was my heart that was calling me back to Africa.

Roger, with characteristic generosity, fully supported my proposed study. In fact he went so far as to offer me, again, a free vehicle and tracker for the two years I'd be living at Humani. Being the recipient of his benevolence, as well as my scholarship and the advice of two awesome supervisors, was humbling. It was too good to be true. Unfortunately, it really was.

A month after I started my doctorate the violent war-veteran invasions of white-owned farms in Zimbabwe exploded across the world's television screens. I can honestly say that I was unperturbed, because I knew how the media blew things out of proportion where Africa was concerned. I tried to placate my worried supervisors and family – assuring them it would all blow over. Meanwhile, thousands of farms were being occupied, farmers killed, their wives raped, houses burned to the ground and countless Shona tortured and beaten to death for working for whites. Zimbabwe went from being one of the best known and loved tourist spots to being a place blacklisted for foreign visitors. Still, I wasn't daunted. I simply refused to believe it.

The president, Robert Mugabe, openly declared all white people 'enemies of the state'. Zimbabwe had been independent of British rule for twenty-five years, during which time Mugabe had not stepped down as leader. From the safety of my house in Brisbane, I watched his contorted face raving that Britain must not tell him what to do, that no matter what Tony Blair said, he, the leader of Zimbabwe, must resist.

However, I was being told from every angle to seriously reconsider my options, so I phoned Humani to ask the Whittalls if everything was okay there. I reassured myself that what, according to the international news, was happening at one end of the country might not be occurring throughout the whole of Zimbabwe. It was a relief to hear Anne's warm voice, with her lovely Zimbabwean accent. She told me that there were indeed war veterans at Humani, but that everything was fine and I should still come. I didn't know it at the time, but their phones were being tapped, so there was a limit to what she could say.

Still, with my supervisors urging me not to go to Zimbabwe yet and my family literally worrying themselves sick, I decided to fly into South Africa first instead. Johan du Toit at the Mammal Research Institute kindly invited me to base myself in Pretoria until I could decide whether I should change my study site. He was watching the situation in Zimbabwe with even more concern than I was: he had family there.

As soon as I decided not to fly directly to Zimbabwe, another door opened, ever so slightly, just enough for me to see what might be on the other side. First, I met a fascinating couple who had worked on elephants in Zimbabwe and now lived on the Sunshine Coast of Queensland. Sally and Jeremy Henderson asked me if I'd ever heard of the black-faced impala of Namibia and showed me pictures of this endangered species taken on their holiday in Etosha National Park. I had read what was then the only existing international scientific paper on the black-faced impala, and had written to the American authors, but had heard nothing back from them. My heart wasn't really in it, though. I still hoped that the problems in Zimbabwe would fizzle out and allow me to do my fieldwork there. I didn't know the first thing about Namibia and it seemed a long way away from Zimbabwe, where I had a surrogate family and friends.

As I flew to South Africa I felt like a piece of driftwood, plummeting downstream, running into banks and boulders, catapulting and rebounding in a haphazard path to an unknown destination. I tried to embrace the unknown, knowing that I would be wiser for the journey, but the lack of direction in my usually carefully planned life was very unsettling.

My plan was to spend my first few days in Johannesburg with friends of a friend. The de Kleijn family was made up of Tineke, an extraordinary woman full of energy and wisdom, her peaceful accountant husband, Renis, and their three teenage and twenty-something accountant and future accountant sons. Tineke, being surrounded by male accountants, seemed happy to have some female company, and they generously welcomed me into their barbed-wire-enclosed home as if I were a long-lost daughter.

Within days of being there my body was taken over by a horrendous virus. My physical state always tends to reflect my emotional one, as I think it does for all of us. My anxiety in the build-up to arriving in South Africa, the unknown status of my future at Humani and the sad goodbyes to my friends and family had taken their toll. My unspoken worries liberated themselves in illness. From my neck to my feet, everything ached. There was a stabbing pain in my lungs and my throat and for about four days I hardly got out of bed. I had never been so sick before. Usually I don't let being ill get me down. But now, far from home and feeling like hell, without direction or the capacity to make things happen any faster than they were, I felt utterly depressed.

Although I was a virtual stranger, Tineke treated me like one of the family. She was resolute in her attempts to make me well and dragged me to the doctor, who diagnosed me with a severe case of tonsillitis and a pulled lung muscle from coughing. Dosed up on painkillers, I almost felt like a million dollars! I rang home to let them know I was still alive, downplaying the fact that I'd been bedridden and semiconscious for almost a week. Mum told me that the Australian Government had issued a general warning to all Australian residents not to go to Zimbabwe, and especially not to commercial farms owned by white people. I phoned Humani. I had to know what was going on.

'Hi, Sarah! It's Tam. How's everything going there?'

'Hello, Battleaxe. Everything's fine. How are you?' Sarah's voice crackled over the phone lines, another conversation in Shona echoing in the background from overlapping lines.

'I'm good... I'm still in Jo'burg. So have the war vets come as far south as Humani yet?'

'Oh no, Tam. Everything's fine here.'

'Really? So no worries at all?' I asked, surprised.

'Actually, Tam, we can't talk about this.' Sarah's voice was strangely urgent and commanding.

Then we were cut off. I suddenly realised that the lines were being tapped. Anything that the Whittalls said on the phone would be recorded and used against them. It was another bad sign. My trip to Zimbabwe had been delayed, and then a shocking virus had stopped me going even as far as Pretoria. It was as if someone or something was trying to tell me that I shouldn't go back to Zimbabwe.

I'd missed my lift to Pretoria, so Tineke kindly offered to drive me there. After she had dropped me off at the University of Pretoria, I walked into the daunting matrix of tall buildings feeling very out of place. I wandered around aimlessly for a while, unable to understand any of the Afrikaans signs, then stopped out the front of a building and asked someone where the zoology building was. Coincidentally (or perhaps, not at all), it *was* the zoology building! It was as though I'd been summoned.

'Tammie, you're alive,' Johan du Toit said when I found his office.

'Only just,' I smiled.

It was hard to believe I was finally there. Johan was professional and businesslike, but there was a kindness behind his dark brown eyes. He looked too young to be a professor, but the fact that he was spoke of his success in the field of conservation biology. He really did seem to empathise with my situation and encouraged me to look at another study site in South Africa, Venetia. Pulling out a map, he showed me the location of Venetia, a property owned by de Beers, the multimillion-dollar diamond-mining company. Johan told me that Venetia, located on the Limpopo River in northern South Africa, was similar to the Save Valley Conservancy in that it was dominated by mopane trees and had lots of impala. This appealed to me.

'It's close to us here at the MRI,' Johan went on, keen for me to give it a shot, 'and we could knock out a project proposal in no time.'

In the back of my mind, I couldn't rule out the possibility of working on the endangered black-faced impala. Johan had been less enthusiastic about Namibia, perhaps because he didn't know as much about it and couldn't offer me as much assistance. Every time I left his office he would say 'good luck' and even though I know he meant well, it seemed as if he was really saying 'you're on your own with this, kid'.

Then, out of the blue, I had a reply from the American authors of the one and only international paper on black-faced impala. Aron Rothstein and Wendy Green had studied the subspecies in Namibia for eight years. I held my breath as I read Aron's letter. It was a turning point.

He explained that there was a great need for conservation-oriented research on the subspecies. There were only three thousand black-faced impala left, perhaps less, because the black-faced impala were interbreeding with the more abundant common impala on Namibian game farms. He gave me a list of people to talk to in Namibia's Ministry of Environment and Tourism and encouraged me to go for it. This was it, I just knew it.

An Aussie friend from university, Elisa, pitched up out of the blue needing a floor to camp on while her boyfriend went to volunteer on a project that had something to do with buffalo testes. Els was at a loose end, so we decided to embark on an Aussie chicks' road trip that would take us from Pretoria in South Africa, across the donkey-riddled Trans-Kalahari Highway in Botswana and end up in the Etosha National Park in Namibia. The journey to Etosha would take us three full days of driving from dawn till dusk.

In the days before we left, I received encouraging emails from scientists in the Namibian Ministry of Environment and Tourism. I was soaring on hope and positive energy when I phoned the Etosha Ecological Institute to ask if we could stay at the park's research camp. I spoke to the acting head of the park, Peter Erb.

He was terse and authoritative with me and said in no uncertain terms that we could *not* stay in the research camp because I didn't have a research permit. I explained that I couldn't apply for a permit until I had done a feasibility study in Etosha, but after a long, painful pause, Peter said it was not his problem. Ouch. After I'd clearly pissed him off – not a good thing considering he was in charge of Etosha – I decided it would be diplomatic to write him a full introductory email, complete with my original study proposal for Zimbabwe. It was fortuitous that I did, as it probably saved us from being kicked out of the park.

We soared across the Kalahari Desert in our white Chico Golf, Els and I each taking turns to drive, the golden winter grasses forming a mosaic of sharp tufts on endless undulating rose-coloured sands. A sprinkling of tall acacia trees, some of them hung heavily with pods, dotted the landscape as we flew along the excellent bitumen highway. I had breathed a sigh of relief after we'd crossed the border out of South Africa – I'd grown accustomed to always being a little tense there. It was good to be back in rural Africa, where towns were villages and where tsama melons and wild animal dung lined the roads rather than litter. Friendly people in donkey carts smiled and waved as we passed.

Botswana is a prosperous country. The national currency, pula, means rain and is stronger than the South African rand. Botswana got rich on diamonds, agriculture, cattle and a thriving tourism industry and, as a result, its rural people are generally better off than those from other African countries. As we drove along the national highway, dodging psychopathic donkeys and mad cows, we were treated to magical sightings of hartebeest, ostrich, jackals, vultures and colourful birds of every description. We cruised across the vegetated, sandy Kalahari where wild and domestic animals roamed freely, eking out an existence in this extraordinarily dry environment. As the sun peeped its head out from behind the illuminated horizon, its radiant, white rays like ghostly fingers reaching for the new day, I felt nature reminding me of my small place in the big picture. After all the tension that had led to this moment, I knew I was on the right path. I felt the warm glow of the sun on the back of my neck like a kind embrace, and goosebumps prickled my arms.

The first person I met in Namibia was John Coetzee, who was then the black-faced impala coordinator for the Namibian Ministry of Environment and Tourism. Els and I arrived in the capital city of Namibia, Windhoek, on a Friday afternoon. When I phoned

John, he told us to meet him at the pub. He said that he would notice us because we would probably be the only women sitting alone together in the pub, which turned out to be true. Over several beers he explained that there was once thought to have been tens of thousands of black-faced impala in the region formerly known as Kaokoland, now the Kunene region, in north-west Namibia. Now, as a result of poaching during the war for independence, competition with the livestock of the Himba people and severe droughts, there were only a few hundred remaining in their historical range. He said all this while rubbing his hairy leg against mine and trying to look down Els's shirt.

The next day we headed north on the five-hour drive to Etosha. One of the first animals we saw when we drove into the park was a black-faced impala.

'What a great omen!' Els cried, as excited as I was.

Soon we arrived at Okaukuejo (pronounced Ok-ah-koo-you), the central camp in Etosha, where the Etosha Ecological Institute is located. A tall Germanic lookout tower surveyed a random assemblage of single-storey brick buildings and rondavels with thatched roofs. To my surprise, Okaukuejo was like a small town. It had a couple of shops, a fuel station, a pool and a bar. But I'd forgotten what day of the week it was. We'd arrived at eight on a Sunday morning; the institute was closed and there was little sign of life save for tourists packing up camps and heading out for a morning of game viewing.

In the staff quarters, located behind the tourist camp, all the houses, cramped in tightly beside each other, were closed up, their windows locked fast. Driving slowly along the wide dirt tracks, I finally found a man who appeared to be watering a threadbare lawn. I asked him to point me in the right direction. I'd been advised by John Coetzee to advise Nigel Berriman, one of the institute's three research technicians, upon my arrival. He would help me get started and show me around, John said. Although all the letters I'd received from scientists in the ministry had been encouraging, I got the feeling that I wasn't to expect any special treatment. It seemed as though some of them were expecting the worst of me even before I started, the sins of past foreign researchers a slight on me even before I had arrived. I wondered what kind of bees' nest I was hurling myself into.

I wound my way through the maze of government housing to a small block of identical flats. I'd been directed to a flat that was surrounded by a kaleidoscopic display of cacti and aloes. Nestled into rock formations were green cacti with thick wispy hair; giant smooth aloes shaped like rose petals in red, yellow and green; tall Mexican-looking cacti covered in long, sharp spikes – a mind-boggling array of all things prickly and resilient.

I knocked gently on the wooden door. No one answered. I knocked harder.

'Okay, okay!' a raspy voice echoed from within the fortress. 'Hold your horses, I'm coming!'

A phlegmy smoker's cough shook the walls and I heard someone spit grotesquely. A key fiddled in the lock and the door creaked open, revealing Nigel in his white, baggy underwear, wiping the sleep out of his eyes.

'What is – ?' he muttered, only then opening his eyes, then slammed shut the door when he saw me standing there. 'Geez! Just wait there, girl, till I put my pants on!'

I heard cupboards opening and doors slamming, things being thrown around the room.

'Did I come at a bad time?' I called from outside, raising my eyebrows at Els.

'Well I wasn't expecting visitors this early on a Sunday morning, but seeing as you're here you may as well come and have a cuppa tea. You can come in now. I'm decent.'

The claustrophobic flat smelt of last night's whiskey, day-old cigarettes and dust. I stifled a sneeze. Overflowing ashtrays littered the arms of old torn leather chairs and spilled onto the concrete floor.

'You have a lovely place here,' I commented.

Books of every description fell upon each other in disarray. An old record player was hidden under layers of dust. A painting of a bull kudu with majestic spiral horns hung slightly lopsided on the wall.

'Ja well, even after twenty years of government service this is the best they give an old man,' he called from the kitchen.

I could hear the kettle boiling.

'Still, I'm not complaining,' he continued, 'long as I got my cup o' five roses tea in the morning and a pack o' smokes.'

Nigel emerged from the kitchen with two cups in his hands, or should I say two *jugs*. Dark, strong tea steamed from huge beer mugs and small lumps of powdered milk floated on the top of the dark brew. As he put the mugs down on the coffee table, a little of the warm liquid spilled onto the glass, merging with dust to form a kind of dust-fringed amoeba.

'Thank you,' we smiled, then almost choked as we sipped the sweet concoction.

'Hmmm...perfect,' I said, thinking that I was getting my sugar quota for the month.

He sat back in an armchair opposite mine, took a large swig of his tea and, for the first time, looked at me. The leather creaked as he shifted his weight. He adjusted his spectacles so they sat higher up on his nose.

'So. Who are you exactly and to what do I owe the pleasure?' He was clearly amused.

'I'm here to do a project on black-faced impala. Well, I'm looking at whether it's feasible really.'

'And how long will you be here for, Tammie?' he asked, swallowing a mouthful of tea.

'Two years I reckon, if it works out.'

Nigel smiled. 'Two years? That's a good length of time. You'll need every last bit of that if you want to understand the black-faced impala.'

I nodded.

'Well, now that you're here I guess you'll need a place to stay. I'll take you over to the research camp in a minute. 'Fraid you'll have to wait till I've had at least two cups of tea first, being a Sunday an' all.'

At least four elephant-sized cups of tea and a desperately full bladder later, we walked over to the Okaukuejo research camp. With four caravans and a couple of large tents, the camp could sleep up to a dozen people comfortably. It had a communal kitchen with a concrete floor and a tin roof, which had been converted from old horse stables. The toilets were identifiable by the acrid smell wafting from a sewerage system that was obviously not working. Nigel had been working on it, he said, but no matter what he did, he just couldn't work out why it stunk like that. I'd get used to it, he assured me.

My caravan, like all the others, was an ancient square box that glinted metallically in the sun. Inside, only one of the three fluorescent lights, draped in spiders' webs, worked. Two tiny single beds, with just enough room for two very small people to sleep in, squished in at one end, and beside them, two warped wooden cupboards were built in. Gecko droppings and spider webs adorned the floors and walls. There were two push-up windows beside the beds, both locked shut. A small aluminium washbasin, stained with white streaks from the calcium in the water, was at the other end. I turned the tap on, but nothing more than a grumpy splutter and a defiant moan came out. Nigel was tentatively opening each of the cupboards and peering inside.

'Just checking,' he said. 'You never know when a zebra snake might've wandered in.'

Suddenly he shouted, 'Aarrgh!' and jumped back.

Els and I bolted outside, crashing past a cupboard door in the attempt.

Nigel leaned out the door, a cheeky grin on his face.

'Just checking your reflexes,' he chuckled.

Over the next few days Els and I watched elephants bathing, black rhinos drinking, black-faced impala rams rutting and met some of the people who would come to be my surrogate family over the next couple of years. Watching my first sunset in Etosha was like hurling myself head-first into hell. It was magnificent. The sky was on fire. Piercing shards of silver light tore towards the heavens through myriad hues of red, from damask rose through to burning maroon. Raging winds had turned the sun to a blazing globe the colour of blood. Dust from the pan was flung this way and that, forming towering, twirling dust devils blown at the command of the wind.

Aaah, I thought, I have arrived. In Etosha, I felt as though I was at last in a place that I could grow to love as much as I loved Humani.

OUT OF DISASTER AND INTO THE DESERT

Before I could officially begin my research in Etosha, I had to learn how to wait once again, this time until the Ministry of Environment and Tourism approved my official research permit. I returned to the Mammal Research Institute in South Africa to finalise my proposal and kill some time, but I'd only been there a week before I received an urgent two-line email from John Coetzee. *I'm going to Kaokoland in a week. Get your arse back to Namibia.* He invited me to join him on a trip to the remote Kunene region to look at the historic range of the black-faced impala and offered for me to stay at his flat in Windhoek before we set off. Els had carried on with her travels in South Africa, so, on my own this time, I excitedly boarded a plane to Namibia. I was amazed at how well things were flowing, and even John's slightly sleazy ways couldn't stop me jumping at the opportunity.

I woke up on my first morning back in Windhoek after an uncomfortable night's sleep on John's black vinyl couch with the warm, wet sensation of his dog's rough tongue on my face. My lungs felt like lead due to all of the cigarette smoke trapped in the small room. I looked up to his bookshelf to see the strange juxtaposition of a Bible perched next to a book entitled *Foods for Fabulous Sex*. I forced myself to be appreciative towards John for giving me a place to stay, but there was something very unsettling about the way he went to the toilet with the door wide open, and I just couldn't get used to the shower, which was located *in* the kitchen next to the fridge, with only a see-through curtain for privacy.

By day I joined John in his smoky office and read up on what little information the ministry library had on black-faced impala. Paint peeled from the off-white office walls and old carpet curled up where it had been ripped, revealing dirty concrete underneath. John hid behind his computer screen and the books and files that were piled up haphazardly on top of his cluttered desk.

He encouraged me to buy the groceries for our trip and some camping gear, which I did, and he said that we'd leave as soon as his boss gave him clearance. So, a long week later, when his boss informed him that he hadn't provided enough evidence of preparation for his trip for her to give him permission, I was in a state of shock. I felt as though I'd come back for nothing except a potential case of lung cancer. I wondered what the hell I was going to do with myself, a foreigner in a strange land where I knew almost no one, while I waited for my research permit to come through. Loneliness birds circled. They were vultures.

Thankfully, just when things really couldn't get much worse, I was saved by two fabulous Aussie blokes and a sympathetic German woman who took me under their wing and gave me an alternative plan of action. Dr Keith Leggett and Julian and Steph Fennessy were working on desert-dwelling elephants and giraffes, vegetation and indigenous communities in the Kunene region for the Desert Research Foundation of Namibia. Finding them was like discovering guardian angels in the desert and, as Aussies tend to do overseas, they adopted me like a long-lost mate. Steph let me sleep on her floor and Keith invited me to join their team on a trip they'd planned to Kaokoland just a week later. I couldn't believe my luck.

Little did I know that some of the people I would meet on this trip to the wide open spaces of Namibia's north-west region would remain my friends for many years to come. Ironically, Keith had done his PhD on aquatic ecology in the Sydney Harbour area, but had later developed a fascination for elephants. The son of a farmer in New South Wales, Keith was a fit forty year old with a propensity for saying exactly what he thought. With his thick Australian accent and booming laugh there was no mistaking his origins, but after more than a decade in Africa, it was obvious that he felt at home in Namibia.

He explained that the purpose of the trip was to liaise with the people in the communal areas to help them understand how forming a community-based conservancy and establishing a tourist or hunting operation could improve their quality of life. It was part of national efforts to promote the conservation of wildlife outside national parks and symbiotically to reduce poverty by devolving natural resource management down to the village level. In effect, this meant giving wildlife an economic value to poor, rural people, and offering them a way to make money out of their natural assets.

In Africa you can't think about wildlife conservation without thinking about people. The philosophy of community-based natural resource management really appealed to me. I had always understood that it was very hard for people with hungry bellies to give a damn about conserving wildlife. After all, Africa's people are a natural part of its environment and their children are the custodians of the land. Why not empower Africans to manage their wildlife by giving them financial incentives to do so?

At every community we visited each meeting went on for several days as people stood up to have their say. I learned that Namibia is culturally diverse: it has twelve official ethnic groups and each has its own language. We lived on *pap* (or *sadza* as it is called in Zimbabwe), a bland but filling carbohydrate-rich, corn-based meal that looks like mashed potato. At each meeting a goat was slaughtered to go with the pap. These people ate pap every day of their lives as a staple food. I found it a satisfying and tasty meal combined with tender goat meat. The odd piece of goat's hair I picked out of my teeth added the necessary fibre component to my diet.

The area around the small dustbowl town of Sesfontein was dominated by two tribes: the Hereros and the Damaras. The Damaras spoke in clicks, much like the San Bushmen, while the Hereros were obvious by their clothing, which reflected both the German colonial era and the Herero's devotion to cows. The Germans were the first Europeans to colonise Namibia, formerly South West Africa, in the late nineteenth century, a process that involved the genocide of perhaps three-quarters of the Herero population, a bloodbath of about sixty thousand people.

In 1948 the country came under South African rule, which imposed apartheid, which literally means 'separation'. Homelands were created for the different groups, such as Damaraland for the Damara people and Bushmanland for the San, although these areas did not represent the true territories of these people. The Namibian dollar is still equivalent to the South African rand and Namibia continues to depend on South Africa for imported goods, hence an import duty of fifteen percent on just about everything you buy. Thankfully, though, apartheid is long gone. In 1990 the South African government was overthrown after a bush war lasting almost twenty-five years and Namibia gained its independence under the rule of Owambo President Sam

Nujoma. After a stable reign of over a decade, Nujoma handed over his party's leadership to Namibia's current president, Hifikepunye Pohamba at the end of 2004.

Although German Namibians are now a minority, their influence is still visible. Even today, Herero women wear embellished dresses with puffed-out skirts similar to those worn by the early German missionaries and settlers. On their heads they wear a distinctive hat with long curls swept up on each side of it, intended to resemble cow horns. I have always been amazed that Herero women determinedly wear this 'traditional' attire despite the heat, long after German rule has ended. Tradition is an important part of life in Namibia, and especially so in the rural areas. It felt like taking a step back in time.

On one occasion, when Keith and Julian were looking for a local chief, an old Herero woman brought out two chairs beside her hut. Keith sat down and Julian, who was about my age, gestured for me to sit on the other one.

'No. You sit,' Keith told Julian hurriedly, before explaining to me that it was Herero custom for men to sit if there weren't enough seats for everyone.

The gnarled old woman then brought out a rusty petrol can for me. She directed a gappy smile at me and nodded solemnly. Once again I was aware of woman's subservient role in African cultures. It simply wasn't questioned. It was tradition.

Of course, as I was hanging out with Australians, this time was a delightful mixture of indigenous Namibian traditions and uninhibited Aussie humour. During one meeting Julian, in his characteristic laconic style, pointed to an obese man sitting in front of us. The man's shorts hung low on his backside, revealing the deep crevasse of his butt crack between rolls of fat. He must have been someone fairly important because being overweight is a symbol of wealth in much of Africa, where people struggle to meet global nutritional standards.

I tried not to laugh as Julian gestured towards the man's butt crack and whispered, 'You could fit a stubby in there.'

In north-west Namibia wide blue skies hung over a stark landscape so primitive and isolated that we could have been on Mars. Spectacular cliffs towered over enormous gorges that made us feel like ants. Wide, meandering, dry riverbeds lined with arid-adapted vegetation carved through desert landscapes of angry, red rocks. It felt like an old land and it was. Namibia's deserts are among the most ancient in the world. The rich, raw colours of the earth formed a natural backdrop to villages, smoking campfires and rattling donkey carts packed full of people. Suicidal goats blasted onto the wide dirt road, hell-bent on securing their demise via an appointment with our bullbar. All around us was dust. I woke up each day with its earthy taste on my dry tongue.

At night we sat and talked by the campfire before laying out our bedrolls and sleeping under the stars. Being the end of June, the height of Namibia's winter, the days were warm but the desert night temperatures dropped so dramatically that a down sleeping bag was barely enough to keep out the chill. As always, I slept lightly in the bush. Admittedly, the whoops of spotted hyaenas hardly helped me sleep soundly, their eerie cry echoing through the night like evil spirits. I recalled stories I'd read of hyaenas dragging people out of their beds by their heads. As it turned out, I needn't have worried about the wild animals, only the domestic ones.

One evening, at a community campsite called Omarumba, a rabid dog ran through our camp, dashed under my legs and almost knocked me off my chair. The dog, a scrawny, skeletal flash of fur in the moonlight, attacked the small Jack Russell of the Afrikaans camp manager. When the man tried to get the dog off his beloved pet, it attacked him too, biting into his arm. In the chaos of chairs falling over, dogs barking ferociously, men yelling and pots and pans toppling off the table, no one could react quickly enough to help. Frothing at the mouth, the dog clearly had rabies and had become psychotic. Suddenly I heard a hard thump and the growling stopped.

The man had pulled the dog's teeth out of his arm and thrown the animal at the wall. It was a brutal end, but inevitable. Both the man and his pet dog had to go to Windhoek the next day for rabies injections. We discovered the next morning that this one rabid dog had killed five other dogs in the nearby village. We were very lucky that it hadn't bitten any more of us or else we all would have been on our way back to Windhoek for rabies injections into our stomachs.

I woke up one morning to find an old Herero man peering down at me in the pre-dawn light. Foggy-eyed, I peeped out from under my sleeping bag.

'You wanted souvenirs?' Keith called over to me, smiling. 'You got 'em.'

He was already up and poking the previous night's campfire into action for some much needed warmth on the freezing morning. We were camped beside a dry riverbed under the protective canopy of some large camel-thorn trees. On the sand beside me, the grinning old man laid a sack of bows and arrows and three gourds made of the wood of commiphora trees that had been carved with hot wire into lively triangular shapes. I sat up on my bedroll and bargained, bleary-eyed, in bed.

'Tea and sugar?' the old man said, which seemed to be about the only English he understood, indicating that he didn't want to be paid in money but in food.

To my surprise, Keith came over with a big bag of sugar and Five Roses tea bags as I was making sign language with the old man and bought the lot! The old fellow was smiling profusely as he left, having sold all of his wares before the sun had even come up. I knew I'd have another chance to buy some souvenirs, so I wasn't worried that Keith had bought everything.

But then, out of the blue, he put a bow and arrow set, plus one of the gourds, on top of my sleeping bag, and said, 'Happy birthday, darlin'.'

He dashed away before I could even say thanks. It really was my birthday – I was astounded that Keith had remembered.

I was with Keith when we heard on the BBC news that Mugabe had won the Zimbabwean elections. Neither of us was surprised, with the prevalent rumours that the elections had been rigged and talk of ballot boxes being found floating down rivers, but we knew that it meant closure on any future we may have had in Zimbabwe. We couldn't work in a country whose president considered white people enemies of the state. The writing was on the wall, but I still couldn't quite believe it. Before I could turn my back on Zimbabwe for my PhD, I felt compelled to return to Humani. I had to be sure.

◆◆

'Tam! Burn it!' Sarah exclaimed urgently, her eyes wild with fear. 'You have to burn every last page of your diary or Mugabe will kill you!'

Panicking, I started ripping pages out of my diary, removing anything that had a reference to Zimbabwe's president or the war vets or anything vaguely political that could land me in hot water.

'He is coming here now, Tam. At any moment he could be here!' Sarah's terrified words rang in my ears as fear contorted my stomach. 'Come on. Hurry!'

OUT OF DISASTER AND INTO THE DESERT

As I ripped out each page of my diary, I felt as though I was tearing shreds of my soul from my body. Sarah lit a flame to each piece of paper as I passed them over. The fire ate through the scribbled pages and reduced each one to ash, each page of evidence burned and buried, ending an era.

I woke with the vivid smell of smoke in my nostrils, feeling tense and unsettled. I was snuggled under a flowery duvet in Sarah's guest room. I had been dreaming. Was it a premonition or a warning to be more careful about what I said while I was in Zimbabwe? Things were different here now. Were the burning pages of my diary symbolic of the end of my time at Humani? I didn't know. All that was clear was that I no longer felt safe here.

Until now, poaching had never been a big problem at Humani. Roger's anti-poaching team patrolled the entire farm regularly and they only occasionally encountered a snare. With the instability in Zimbabwe, that had changed. Being in one of the driest and least arable parts of the country, the Whittalls had always hoped that Humani would be safe from what was being described as a free-for-all land grab. The government had talked about 'expropriating' white farmers from their land for years. Although they knew that nothing was certain in Zimbabwe, I don't think the Whittalls ever thought it would actually happen. And when it started, the Zimbos I knew didn't worry much about it. Life went on. The Whittalls had survived the war of independence. They could survive this. With unfailing resilience, they clung onto these hopes because they were all they had.

Sarah pointed out some war vets to me as we drove past their encampment on our way to her parents' house. They were scowling young men who looked to be about twenty-one – and they claimed to be venerable veterans of a war that ended twenty years ago. They must have been the youngest freedom fighters in world history.

The Whittalls knew that sooner or later men claiming to be war veterans would come and try to set up homes at Humani. I think they were even prepared to give up some of their land. It was happening all over the country in a dictator-led strategy to find land for the ever-growing population. Rather than sell one of his mansions to feed his people, Mugabe built his wife another palace and told white farmers to stop being greedy. Like scapegoats they were blamed for the poverty and the landlessness, the hungry mouths and even HIV-AIDS. According to the propaganda, AIDS was a disease that allegedly had been created and spread by white men in order to obliterate the black races. And there was little doubt that many of the rural, uneducated people in the impoverished communal settlements considered there to be some truth in it. When you have hungry children, all rational thought subsides.

Initially the Whittalls considered themselves to be quite lucky. In other parts of the country violence had escalated rapidly. A couple I heard about who lived on a tobacco farm in the prime agricultural land north of Harare had been slaughtered in their beds while their two terrified children hid in a cupboard. The perpetrator was supposed to be negotiating the handover of the farm the next morning, a deal that these particular farmers were prepared to make. They knew that to stay would be to die and they had children to think about. So they and their two daughters had packed their bags the night before and had prepared themselves to leave their homes with nothing but the contents of their suitcases. Instead, the negotiator arrived in the night and ensured that only one side of the story would be heard at the handover the next day.

At first there were only a few people camping at Humani, but soon there were hundreds. Collecting wood and grass as they wished, building their huts wherever they chose, these people created settlements at their will. It was the president's wish, they scoffed. They were entitled to the land. It was theirs, they said, and told the Whittalls to get off. Many of them weren't veterans at all but opportunistic poachers taking advantage of the chaos. Even amongst the war vets themselves there was disgruntlement and disagreement, with different factions turning on each other. Unlike Humani's Shona staff, the people in the war-vet settlement growled at us when we drove past. There were no friendly waves or smiles, for which the Shona people had always been famous. White people were considered intruders on their territory.

Humani's staff had never been very accepting of outsiders, as I'd seen with Mr Hunde the school principal. They'd worked hard to secure their jobs on the farm and to keep them. I felt that if Humani belonged to anyone other than the Whittalls, it was to them and not to these people who had come from all across the country to claim free land. Some of the staff had worked for the Whittalls for over twenty years and had raised families who also worked at Humani. Now, without invitation or agreement, outsiders had arrived in the dozens and the Shona people of Humani found themselves being harassed for being traitors in the new, 'improved' Zimbabwe. They must have been afraid, yet they stood firm.

First and foremost, it was the animals that paid the price for the seemingly inexhaustible hunger for land. Poachers were regularly apprehended by the game-scouts, caught in the act of laying snares or removing carcasses from the bush. The police would be alerted and come out to the farm to interview the culprits. Names would be taken, political affiliations noted and the police officers would umm and aah for several hours before the poachers were escorted to the Chiredzi police station. After a night in the lock-up, the same faces would be back on Humani, camped at the war vet settlement and happily barbecuing their poached meat.

It became an enormous job to remove snares from the property, only to have them replaced by new ones almost immediately. The game-scouts were walking many kilometres every day, returning with pile upon pile of wire traps. I felt sick at the thought of how much fence line had been destroyed to create the traps and the numbers of creatures whose lives had been taken in their vicious clutches. It was becoming an almost impossible task to prevent the game-scouts attacking the war vets who were making their lives unbearable.

One time Roger had to dump the scouts at a remote patch in the middle of the bush to let their tempers cool down. They'd found so many animals in snares, half of which had rotted by the time they'd found them, the meat wasted after an agonising death. When there wasn't enough food to go around, this blatant waste disgusted the scouts. They couldn't understand why anyone would cause such destruction. Like a simmering volcano, the animosity between the people of Humani and the outsiders was building. It had a will of its own. Roger knew he would only be able to separate them for so long.

While I was there, a gang of about fifty war veterans took one of the game-scouts against his will and handcuffed him. They kept him without food or water all day, whipping him repeatedly, demanding to be told the whereabouts of the head scout, John, so that he could be prosecuted – the tribal way. When the man didn't yield by the afternoon, despite repeated torture, ten more of the scouts were abducted and lashed severely with hippo-hide whips. Due to the size of the gang, there was nothing Roger could do to prevent the torture and the police didn't arrive till late in the afternoon, by which time it was too late. The game-scouts were being punished for their loyalty to the Whittalls and for their role in apprehending the poachers who were snaring animals on the farm.

OUT OF DISASTER AND INTO THE DESERT

The price of a loaf of bread had risen to several thousand Zimbabwean dollars, more than most people earned in a week. People couldn't afford to buy basic household items like sugar and toilet paper. The Whittalls supplemented the rations of their staff with vegetables from the garden and meat, but in truth they had begun to ration themselves. The number of tourists and safari hunters had dwindled, and with it Humani's primary source of income.

Malnutrition sent the infant mortality rate even higher than it had been before. At least a quarter of the population was thought to have AIDS. Formerly productive croplands that had fed the population went to ruin as they were taken over by people without agricultural skills, leading to widespread starvation. Zimbabwe was kicked out of the Commonwealth because of its tyrannical president, while its once smiling people battled just to stay alive.

I understood and supported the need for reclamation of land for rural Zimbabweans and for a more even spread of the wealth. But I just couldn't believe how much destruction had to come in the wake of it. It was heartbreaking.

As the animal numbers dwindled at Humani, decimated by illegal hunting, it became harder and harder to snare anything. As a result, snares were left unchecked for weeks at a time, leaving any animals caught in them to die slowly and excruciatingly. The war vets began to stand at the front gates of Roger and Anne's house, waiting for someone to come out so that they could demand meat. They wielded pangas and knives threateningly and there was no choice but to give them what they wanted. Roger would rather that than they obtain their meat by snaring.

The herds of animals were no longer approachable either in the vehicle or on foot. Terrified of the invasive humans that were suddenly persecuting them, they lost their trust in us as a species. And who could blame them? The most one could expect to see now was the tail end of a kudu, its cotton-wool tail flicking before it was absorbed into the protective camouflage of the bush.

Though it saddened me to drive around on the farm, I still did, often seeing animals that had snares wrapped around a leg or a neck. One day I joined Graham Conier, the dark-bearded, stocky chairman of the conservancy, to investigate the horrific scene of a slaughtered elephant. The carcass had been rotting for three weeks and the smell was overwhelming. He'd been a young bull in good condition and was now nothing more than a stinking pile of putrefaction. One of his tusks had been stolen, hacked out. With a metal detector, Graham scanned the elephant's body, looking for a bullet as evidence. Two of the game-scouts were pulling out handfuls of the grotesque, rotten meat with their bare hands so Graham could scan them. It was nauseating.

Although we never found a bullet, the slaughtered bull came to symbolise for me the wildlife persecution that would follow in the conservancy for many more years. It was so sad to think that less than ten years ago these elephants had been translocated to the conservancy from Gonarezhou National Park to protect them from the poaching there.

Later, the stench of rotten elephant meat on our boots, Graham and I drove back to Roger's house. On the way we had to drive past one of the war vet settlements: as we got near to it, we saw that the veterans had deposited a large mopane tree across the road to function as a roadblock. This was a testosterone-inflamed way for them to enforce what they felt was their authority over the white farmers. Half-a-dozen men stood at the block with pangas, ready, it seemed, for war. Unable to drive past the log, Graham was forced to stop.

'Lock your door,' Graham told me grimly.

My heart froze. Anything could happen.

'Yes?' Graham said as the head veteran meandered over to the vehicle.

Graham was grasping the steering wheel so tightly that his knuckles had turned white, and he didn't turn the engine off. I could feel his anxiety, it mirrored my own. I had heard some of the stories about what the vets had done to white women.

The scowling headman, with his maroon beret, glared at us but then commanded his men to move the tree off the road. It was Roger they wanted, not Graham. Still, it wasn't until a long time later that my heart rate returned to normal.

A few days later Roger was abducted. It all began when he was summoned over the radio to Zaharis, the area where most of Humani's crops were grown. Roger drove out to the meeting place, about ten kilometres from his house. When he got there, the men began telling him their demands in commanding tones. Roger listened without reacting. Perhaps to provoke a reaction from him, one of them shouted, 'Why do you whites always go around in vehicles? Why don't you walk like us?'

Roger said calmly, 'Fine. You want to walk. Let's walk.'

And with that, he began to walk back to the house. In typical Roger style, he walked like a man on a mission, powering along despite his rotund belly and gout, the product of years of good safari cooking. The war vets puffed along behind him. By the time they got to his house, the men were thoroughly pissed off that they'd been made to walk so far. So, in retribution, they grabbed Roger, threatened him with their pangas if he dared to leave, and held him captive out the front of his house.

Inside the house my friend Dan, a professional hunter, was radioing back and forth across the conservancy to let people know what was happening. Roger's daughter, Debbie, phoned from her farm north of Harare, in tears at the thought of what might happen to her father. Bizarrely, on the television Roger's son Guy was playing cricket for Zimbabwe. We wondered if Roger wouldn't be more annoyed about missing the cricket match than anything else. Every time someone went outside they told him the cricket score. Roger was known for his stubborn, determined nature, which was why we were worried – we knew he wouldn't bow to the war vets' whims, and this could threaten his life.

Anne, Sarah, Dan, myself and other friends waited nervously in the lounge room; there was no knowing how this might end. Suddenly two war veterans blasted into the room with Roger's older brother, Richard. We all stood up while Richard introduced us to let them know we were not a threat. They wanted to phone the president's office. These guys weren't joking: they wanted orders from Mugabe himself. Anne showed them where the office was, trying desperately to contain her anger at these men who were holding her husband hostage and barging through her home. After they were done – clearly they hadn't got through to Mugabe – they came back to the lounge room and demanded to use a radio to call the police. They believed that the police would support their abduction of Roger and back them up with rifles. Frankly, the way things were going, we didn't know who was on whose side, so anything was possible.

The head war vet picked up a radio and held it against his ear, yelling into the air, 'Allo! Allo!'

Plainly, he'd never used a radio before, which made me wonder how he could be a genuine war veteran. The man was growling at us. He'd lost face once already that day. I tried to avoid eye contact with him, treating him as I would a predator.

The day dragged on agonisingly slowly and the sun began to fade. Eventually the war vets seemed to lose their fire – perhaps they were getting hungry for their evening meal and felt that they had made their point. They let Roger go. We all breathed normally for the first time that day, but no one slept well that night. The truth was, no one knew what was going to happen next. Zimbos had lived with unpredictability their entire lives. Even now, with their family's legacy at stake, I knew the Whittalls wouldn't leave their farm. And five years later, as I write this, they are still there, soldiering on in the face of adversity and turmoil.

For me, although I felt like a traitor, I knew that I couldn't stay at Humani. My study animal, the impala, was being poached to high hell and it was no longer safe to drive around. After I left, a sense of guilt gnawed at me continually; I felt like a deserter. But fate had gripped us all in her mighty claws and decided upon our destinies. Much as it hurt, I had to put Zimbabwe in my past. Namibia was my future.

WHAT IT FEELS LIKE TO BE PREY

In the beginning, still full of turmoil at leaving Humani, I realised how lucky I was in that, unlike most Zimbabweans, I had somewhere else to go.

Etosha. Before I knew the first thing about the place, the sound of the name alone conjured wondrous images of Africa at its most exotic and magical, stirring excitement in the pit of my stomach like a Zulu war cry. Eto-SHA! A word that should be shouted. In the beginning, I never imagined that I would come to call this thirsty, ancient landscape my home, to know her longings like a sister, to feel her pains and joys as her unpredictable seasons threaded through my own.

Xom the Heikom Bushmen christened this place, meaning place of mirages where the earth's skin has been scraped away. It was an appropriate name for the glaring Etosha pan, given by a peaceful nomadic people who once hunted game with bows and poisoned arrows on the vast grassy plains that fringe the monstrous saltpan, and who at the time were themselves being hunted by other tribes. Some also called the pan the Lake of a Mother's Tears, after the unbearable grief of a Heikom mother upon the death of her baby. The Owambo people, now Namibia's dominant ethnic group, gave Etosha its name, which means the place of dry water. An apt contradiction because this was a place where nothing was what it seemed. In Etosha, the heat haze plays tricks with your mind and eyes. You confuse earth with water, sky with saltpan, reality with dreams, predators with prey. Learning to see through the haze is all part of survival in Etosha.

It took a little while for my research project to take shape. Because so little was known about black-faced impala, the government wanted enough basic information from my project to formulate a management strategy for their conservation. With that in mind, we decided to focus the work on habitat preferences, behavioural patterns and factors affecting the survival of black-faced impala after they were translocated to new areas. Half of the global population of black-faced impala, about one thousand five hundred animals, lived in Etosha, so it made sense to base my study there.

And it was in Etosha that my study species for the next three years taught me my first crucial lesson. It's always a good idea to keep an eye on the predators.

It was late morning in my first month at Etosha and I'd been sitting in my *bakkie* at a waterhole named Olifantsbad. Olifantsbad means elephant's bath and, as the name suggests, it attracts great boisterous herds of elephants to drink, swim and play throughout the dry season, from May to December. In the dust and blistering heat, their dry skins parched by the searing sun, the herds rumbled – almost ran – to the water, led by their matriarch, little ones bumbling under adults' tummies, trunks held high, smelling the water. Several would immediately plunge into the muddy water, uninhibited, joyously rolling around in its exquisite liquid embrace. The small ones shrieked with glee; some of the newest additions to the herd were tentative about this strange wet substance at first, dipping a toe or a trunk in hesitantly. Others doused themselves liberally with the cool, muddy water, transforming themselves into chocolate-coated elephants. The large ones trumpeted and squirted great bursts of water over their shoulders and onto their sunburned grey rumps.

Two or three adult bulls, old men in baggy pants, would usually linger at the waterhole in the afternoons, like a boys' club at the bar. They interacted with each other occasionally, trunks touching mouths, leaning against one another's colossal bulk, rough skin rubbing rough skin, but mostly they just stood around watching, waiting, being. The sight of the water and the frolicking females usually induced erections of grandiose proportions in their otherwise lazy demeanours. Their bodies cooled, their thick skins dappled with water, the herds flung clouds of dust over themselves, creating a screen against the sun's biting rays, and evaporated into the bush.

On this particular day, as the elephants were sucked away into the woodland, a bachelor herd of impala rams, their varied ages obvious from the size of their horns, materialised from the trees. Ignoring my car, the handsome rams plodded across the glaring gravel parking area towards the water. Impalas mate during the winter but I had noticed that male black-faced impala in Etosha still chased each other, roaring and flashing their bushy white tails, and fought with each other in clashes of horns throughout much of the year. Perhaps the younger males were testing each other's strength, preparing for the day that they would become dominant males with territories of their own to defend. At any moment a ram could decide that another male was invading his personal space and challenge him to a duel by lowering his head and displaying his lyre-shaped horns. Of course, when it wasn't the rut, most of this behaviour was only play, especially when a male was so young that his horns were nothing more than short, sharp spikes.

Suddenly I was surrounded by males chasing each other around my car as if it wasn't there. They were so close I could have reached out and touched the rippling, short brown fur of their rumps, shiny with sweat. Roaring energetically, their hard hooves sending dust flying up above the white earth, they created such pandemonium that other males arrived to join in the ruckus. It was like being in a pub brawl. The two rams chasing each other were being cheered on by the roaring, head-butting males around them, egging the combatants on by running around them. I spun in my seat, trying to keep my eye on the whorl of grunting rams. It was incredible to be surrounded by impala, watching their natural behaviour, as if, to them, my vehicle was simply a part of the scenery. Unlike the herds of common impala at Humani that Ipheas and I had struggled to stalk and observe, the black-faced impala in Etosha ignored vehicles. They were a behavioural ecologist's dream.

Eventually the rams calmed down and moseyed over to the waterhole. I felt a trickle of sweat snake down my back as the midday heat intensified. I decided to leave and head back to Okaukuejo for lunch. At that moment, a tourist sedan drove in, with the driver waving at me frantically.

'There are five lion!' the man exclaimed in a German accent, wide-eyed and slightly berserk, a common response on seeing lions.

I was less interested in the lions than I was in seeing whether one of my study animals was about to get nailed. Sure, lions are cool, but I've never completely understood the obsession that humans harbour for them. It used to drive me crazy when people would see the official research sticker on my car and ask me time and again, 'Where are the lions?' After all, this oversized pussycat sleeps for twenty hours of every day and I very rarely saw them doing any much more than this. Lions are professional snorers.

From the edges of the woodland I watched as a young, shaggy-maned male and two juvenile lions followed a leading lioness. The four of them stalked towards the water. With low-hung, full bellies, the golden cats were intent on drinking, not hunting. Even before they had materialised, the zebra and impala at the waterhole had been standing on their guard, snorting and looking in the direction from which the lions came. To my surprise, only a few of them ran off. A couple of zebras backed away from the other side of the waterhole, stamping a foot cantankerously or *haw-hawing* in warning. A few of the impalas barked, cutting the air with their alarm call, deep and sharp.

The lions drank, crouched down, lapping at the water's edge, careful not to wet their sensitive paws. They glanced up every so often to monitor the location of the zebras and impalas, but they didn't seem inclined to hunt. The prey animals seemed to know this, standing vigilant but not overly alarmed.

When the lions were finished, they walked up the bank of the waterhole, past the alert zebras and impalas, and past me, sweating in the parking lot. No doubt they were in search of a shady tree, intent on a snooze that would last until the sun went down and the hunting began again.

As they slunk away, some of the impala began to follow them. Standing tall, their necks held high and straight, the delicate antelope walked tentatively towards them, arching their heads to get a good look, as if fascinated by them. The lions were soon lost from my sight, but I could still see the impalas following them, tentatively but purposefully.

At first I didn't understand why the impalas were doing this and I thought they were being pretty stupid! I mean, what kind of a crazy animal follows the predator that wants to eat them?

Did they have a death wish? Was this why they were endangered? It took me a while to realise that by keeping the lions in their sight, the impala were actually preventing a potentially deadly situation in which the lions had the benefit of surprise. Lions are ambush predators. They stalk up as close to their prey as they can get and then make an explosive rush to bring the animal down. Their success depends on the element of surprise. If the lion is detected by its prey, and the distance between them is too great, then the hunt fails. By keeping an eye on the lions, no matter how terrified they may have been, the impala were ensuring that they wouldn't be caught out unexpectedly in the open, vulnerable environment of a waterhole. To live another day in Africa, I thought, you have to be prepared to look your fears in the face.

Sometimes, though, you don't even know you have a fear until you're already facing it.

I'd been in Etosha a couple of months when I realised that to fully understand the issues facing the conservation of black-faced impala, I'd have to look at how they were doing on private land. About a thousand, one-third of the population, lived on commercial farms that operated tourism or trophy-hunting operations. The biggest threat to black-faced impala was, ironically, common impala. This more abundant subspecies had been introduced to Namibian game farms largely from South Africa by farmers wanting to build up their overall animal numbers. Unfortunately, the two subspecies interbred readily, so on farms where rare black-faced impalas were already present they were being wiped out by hybridisation. It was bizarre to think that the subspecies of impala that I had worked on in Zimbabwe was now 'the enemy' in my new project in Namibia.

It was September when I drove across western Etosha, which is off limits to tourists, to visit some farms bordering the park. According to the government records I'd been given, some of these farms had both black-faced and common impalas, which meant that we would probably have to write their populations off as hybrids. After a few days of talking to welcoming German and Afrikaans farmers in haphazard English and driving around their properties, I was returning to the park with some grim findings. Common and black-faced impalas were moving through game fences and mingling easily on several farms. The hybridisation problem was even more widespread than the government had imagined, which meant black-faced impala were possibly more endangered than we'd thought.

As I drove back to the park, it became obvious to me that it was too late to make it all the way back to Okaukuejo that day. It was illegal to drive in the park after dark, a policy designed to give the animals a break from people during the night. I decided to camp at Kaross, the endangered species breeding area in the south-west corner of Etosha. Supposedly Kaross was predator and elephant free, to make it safe for species like black rhino, roan antelope and black-faced impala to breed up for reintroduction elsewhere.

This fenced-in haven is one of the most beautiful parts of Etosha. The rocky dirt track to Zum Haus campsite meanders up and down the rolling topography, past piles of large boulders that form picturesque *kopjes*, into open, sandy savannas dotted with tall acacia trees. As I drove into Kaross for the first time, I felt as though I was Alice entering Wonderland. On the track the recent spoor of a black rhino, dinner-plate sized, and its four-toed footprints marked the animal's path in the sand. A magnificent secretary bird standing on a massive nest at the top of an acacia tree peered down at me as I rounded a bend, and a herd of black-faced impala danced and bounded across the open plain into the protection of cover. I noticed what I thought was the fresh spoor of two spotted hyaenas on the road, recognisable from the slanted edge of the pad marks and the fact that one pad – the front one – was bigger than the other. Kaross seemed to be crawling with animals, but being off-bounds to tourists, they were skittish and not accustomed to vehicles.

For some reason the animals' skittishness made me uneasy. The warden of this part of the park was aware I was camping overnight here, but it was a strange feeling to know that I was out in the open without any prospect of seeing another human until I headed for Okaukuejo the next day. If I got lost or the car broke down, how long would it take for someone to find me?

I pulled up at the small, tin-roofed rondavel overlooking Kaross Fontein waterhole where I would camp. The grey-brick rondavel was built beside a natural formation of large boulders and featured an open-air shower nestled into the rocks, surrounded by a reed wall for privacy. A metal door and small broken glass window were the only openings in the round single-room building. There was nothing inside but owl and gecko droppings and spiders' webs. Perched up high on top of a rise, the views of the surrounding wonderland of *kopjes* and plains from the camp were spectacular.

The wind whipped around me as I got out of the car, further unsettling me with its maniacal fury. When it is windy animals' hearing doesn't function very well. This makes prey animals more vulnerable and they become edgy. The past week had been very hot, so I was wearing shorts and a light singlet top. Now the day had magically turned cold and the wind felt as though it was blowing straight off the Antarctic. I shivered, wishing I'd brought a jumper with me. I began to feel very aware of being alone. To distract myself I checked for animal spoor in the camp. Based on my meagre knowledge of tracks, I could see that zebras, oryx and impalas had made use of the area recently. I'd been told that there were no lion in Kaross, so I forced myself not to worry.

As I made a fire and cooked some baked beans and toast over its comforting flames, the sun dipped below the horizon, stealing away with it the last remnant of daylight with a suddenness that only Africa knows. I huddled close to the fire as the wind continued to blow through the exposed camp, roaring violently as if to vent its anger. Without anyone to talk to, I grew more and more aware of my solitariness. I was surprised to find myself feeling lonely. Here I was in this magnificent environment, not able to share it with anyone. Yet I'd never thought I needed people much and cherished time for myself. I contemplated why I was feeling so tense and I came to the conclusion that perhaps humans aren't meant to be solitary creatures. They say that researchers grow to look and act like the animal they study. I realised that I was more like an impala than I'd thought, gregarious by nature and needing the comfort of others of my kind to feel safe.

At about eight, with nothing else to do, I snuggled into my sleeping bag on my swag inside the small house. The brick walls kept most of the violent wind out, but it still blew in through the cracked window. The cold, like a devilish African spirit, crept in through the cement under my swag's mattress, and no matter what I did I couldn't get warm. Alone in the dark, with an owl screeching as it flew in and out of the window, a bat flying around the roof and hyaenas screaming their harrowing, primeval howl in the distance, I barely slept a wink. I could hear the deep, guttural *humph* of lions too, but I convinced myself that they were outside the park's fences. Still, instinctively, I didn't feel safe.

I'd always considered myself to be fairly brave. I'd spent many nights camping in the bush in Africa and Australia. But I had never been completely alone. So much for my independent, tough girl reputation! Who was I kidding? I was forced to be brutally honest with myself: I felt deeply insecure by myself, so much so that my body refused to relax into sleep. The night went on forever.

WHAT IT FEELS LIKE TO BE PREY

Before the sun even rose, when there was just enough light to see by, I loaded up my car and headed back to Okaukuejo. I had to scrape a layer of ice off the top of my food containers, bag and car windscreen – the night had frozen everything around me.

When I got back to camp, my tin caravan had never felt more like home. And perhaps, after all, my instincts had been telling me something, because I found out the next day that lions had broken through the fence at Kaross and had been prowling around while I'd been trying to sleep.

The one thing that worried me most about living in Etosha was the lack of people my own age to hang out with. I was twenty-two, single and surrounded largely by middle-aged, Afrikaans bachelors who seemed to keep to themselves. I was living my dream, but I longed for some kindred spirits to share it with. Would I lose my social skills and turn into a drivelling bush-mad freak with birds nesting in my hair? I'd fallen on my feet where my research project was concerned, but as far as a social life went, I'd fallen into a dry well in the middle of a desert and my high heels and lipstick were nowhere to be found.

Even though English was the official national language, it was clear that Afrikaans was the tongue most frequently spoken by the many different races in Etosha. Without a basic understanding of it, I felt excluded in Okaukuejo. Even though I'd picked up enough Afrikaans to remain polite in the few months I'd been in Namibia, I quickly learned that fluency was a matter of necessity if I wanted to become a part of this isolated community.

In the Etosha Ecological Institute's tearoom, where every day, the high-ranking staff gathered ritually at ten and three for their half-hour tea breaks, Afrikaans was the language of conversation, though they could all speak English. Isolated by the language barrier, surrounded by burly, middle-aged blokes, spitting and cursing and occasionally glancing at me like I was some weird species they couldn't identify, I could not have felt more like an outsider. I was too young, too foreign and too female to be taken even vaguely seriously by this boys' club. Being a woman had never stopped me doing anything before, and even though I didn't consider myself to be a brilliant scientist, I'd worked hard to get where I was. So I was surprised to find myself a little shaky around the men in Etosha with their gruff and condescending ways. There was no doubt about it: age was making me soft. I began to feel strangely uncertain of myself, as if I had no rights in this esteemed male territory.

The chief warden of Etosha, Coen Karstens, had barely said two words to me since I'd arrived. Well-rounded and pink-faced, the first thing I noticed about this gruff Afrikaner was his bushy grey moustache. When he talked, which didn't seem to be often, it came alive like a dancing ferret on his upper lip. He was in charge of managing the entire park, which seemed to involve an abominable quantity of paperwork, if the pile on his desk was anything to go by. Every time he looked down at it, the ferret wriggled involuntarily. Actually, Coen didn't seem too fearsome to me, only shy. He had kind eyes, but the last thing I wanted was to get on the wrong side of him. One word from him and I was out on my ear. Still, it would have helped if he would just talk to me.

Mind you, in comparison with some of the others in the institute, he was a pleasure. For example, I'd fallen on the wrong side of Peter Erb, the acting chief of research, even before I'd arrived. He was quick to let me know my place within the hierarchy now that I had my research permit. Like many German Namibians, Peter was a serious fellow who did things by the book and to perfection. Keeping foreign scientists like me in line was part of his job.

'Tammie, we are supposed to be running a park here...' Long, tense pause. 'There are *channels*, you know, channels that have been created for the explicit reason of making the system work, channels that must be followed and which lots of people before you have ignored.'

He pushed his glasses up onto his nose, looking past me as if he had much more important things to be worried about than a new researcher in the park. I didn't doubt that was true.

'I'm sorry, Peter. I really don't mean to overstep any boundaries. It's just that if I can't have people in the back of my *bakkie* then it's going to be impossible to conduct game counts properly.'

'Well, I'm afraid that's not my problem.' Long pause. 'You simply have to get *traalies*.'

'*Traalies*?' I had never heard of such things.

'Yes, bars on the back to keep people in…' Another painful pause. '…like we have in the ministry.'

'That's going to cost a fortune…' But I knew he wouldn't bend. 'Well, do you know where I can get it done?'

'No…' Endless painful pause. 'I'm afraid not.'

As far as he was concerned, I was on my own. I wondered what previous foreign researchers had done to piss him off. It was going to take hard evidence to prove my worth here. It seemed that as far as Peter was concerned, I was just another privileged foreigner out for a fun time.

I was quickly learning about the undercurrents in the institute. From what I could gather, the research division didn't listen to the management division and thought they were useless, while management thought research were a bunch of snobbish academics who had no idea how to run parks. On top of that, Afrikaners and German Namibians differed markedly in their approaches. Both cultures had colonised and governed Namibia, then South West Africa, prior to independence. Racial divides, I was learning, still ran deep on some matters, while on some days I'd see them all congregating at the bar like old mates. It was so complicated. And that was only the white tribes!

The Owambos were the president's tribe and they dominated Namibia's population. The push towards empowering formerly disadvantaged tribes known as affirmative action created bitterness between different tribes, often with inexperienced people being pushed into jobs because of their race rather than their ability and credentials. But it had to be done. It was a generation of change and adaptation in Namibia, long after apartheid had removed its evil claws of racial separatism from the country. Most people were glad to see apartheid over, but the aftermath was not an easy era to live in, no matter what tribe you hailed from, because everyone was trying to find their feet on a shifting substrate. It really wasn't a simple issue that could be explained in black and white. As an Aussie, I realised I had the benefit of an outside perspective. If I'd been born in Namibia, my outlook would probably have been coloured by my tribe. And yet, despite the prevailing negative attitudes in Etosha, there were a few who saw the present as a very exciting time, a new start with enormous potential for individuals to make positive changes that would shape the history of Namibia forever. This gave me hope.

'You should learn to speak a bit more of the Dutchie lingo, Tam,' Nigel encouraged me. 'Honest. It'll make your life a lot easier.'

'Yeah, I know. I will. *Ek praat bitje.* I speak a little bit...' I answered tentatively.

Nigel nodded sympathetically. 'It's actually quite an easy language to learn. You just need to learn to spit when you talk.'

That night, over a Tafel lager at the thatch-roofed Okaukuejo bar, Nigel started me on a few more Afrikaans words – starting with the swear words, naturally. Then every day after that he taught me another couple. Each day my vocabulary increased, but

I was still too unsure of the language to speak it in the tearoom. Instead, I spoke only when spoken to, which was probably exactly how the boys' club liked it!

Then, to my great relief, Okaukuejo received a new warden, Shayne Kotting, and a research technician, Birgit, Shayne's wife. They were both about my age and had been transferred from Halali, the neighbouring campsite. With Birgit in the tearoom, being a female didn't feel so odd. She was quiet and shy and exceptional at her job, as well as looking like a supermodel in khaki. Shayne, on the other hand, was anything but shy. With his dark brown hair, black beard and intense, brown eyes, I knew that he was a man not to be messed with. He had a wicked glint in his eye and I couldn't tell if he was truly terrifying or just a lot of fun.

'Matson!'

The command was yelled from Shayne's office as three rangers scurried out carrying automatic rifles, their faces scowling. I entered quickly, feeling as if I should salute with all the khaki and epaulettes around this place.

'Take a seat,' Shayne instructed, 'and do me a favour and close that door behind you before I kill somebody.'

The office was cluttered with metal and wooden cabinets filled with files. At the centre of the cramped space was a large wooden desk on which papers were arranged in tidy piles and stacked in baskets. Shayne appeared to be filling in forms in a rapid, messy scrawl.

Suddenly he threw the pen at the wall and cursed, 'Bloody thing!'

Then, just as quickly, he smiled at me and started laughing. 'Sorry man,' he said. 'Don't worry about me. I've just got a case of Monday-itis.'

'But it's Wednesday.'

'Ach well, here we get Monday-itis any day of the week. It's nothing serious. These damned numb-nuts in head office keep sending me bloody forms to fill in and I'm sick to death of them.'

He pulled open a box under his desk and pulled out a mobile park radio and charger. I signed a bunch of acknowledgement forms, which said something to the effect that if anything happened to the equipment I'd have to pay for it. My call sign was 88, Shayne said. His was 10. Any problems whatsoever and I was to radio him directly, he told me. He seemed genuinely concerned about my safety and I caught a glimmer of softness beneath his hard exterior. Like everyone else here, his tough veneer was a protective shell.

He suddenly smiled and asked, 'So Tammie, what's a girl like you doing in a place like Etosha? Surely you've got better things to be doing with the prime years of your life than hanging out in a hot little hole like this.'

'Well, funny you ask that because I've been wondering the same thing myself,' I answered, and we both laughed.

'Actually, so far I kind of like it here. It's like nowhere else I've ever lived. And it's not as if there's any going back to Zim now with the war vets tearing up the place.'

'But you haven't answered my question,' Shayne probed.

I paused to think about it. What was it exactly that had brought me to this harsh, dry place, fighting for air among a bunch of middle-aged Afrikaans bachelors?

'I guess you could say I'm following a dream.'

'Aha.' Shayne smiled. 'Well in that case, you'll fit in real well here, Tammie, because you must be mad…just like the rest of us!'

Finally, in Shayne, I knew I had found a friend. Birgit was away studying a lot at that time and he appreciated the company as much as I did. Then, out of the blue, an English girl who'd been working on leopards west of Etosha pitched up and asked if she could volunteer for my project for a couple of weeks. Naomi had a Masters in conservation and was overqualified to volunteer, but frankly, I was so desperate for some young female company that I was delighted to have her. There was hope for my social life after all! What started as two weeks volunteering turned into six months; Naomi became a cherished friend and the two of us were so inseparable that we became known as Tweedle Dee and Tweedle Dum: partners in crime.

I hadn't camped out since my night alone at Kaross, so now, with Naomi, I decided to head for Namutoni in the far east of Etosha. The old horse camp was surrounded by a three metre high elephant- and lion-proof fence; there was a large canvas tent in one corner where Nad Brain, the Etosha vet, camped with his wife Ginger and son Kimber when he was working in the area, and a small fireplace at the front with a large pile of firewood.

By day, as we drove around the habitats of the Namutoni area, learning where to find impala, I began acquainting myself with possible routes for impala population counts. Tall tamboti trees with rough bark and distinctive canopies towered above a dense grove of thorny, leafy bushes in which diminutive dikdik antelopes hid. Honeycomb-coloured plains, dotted with herds of wildebeest and springbok, surrounded the edge of the cracked, dry Etosha pan, which was a kind of reef blue from the mixture of minerals and microbes in it. In the dry season the pan became the mirage of an inland ocean, the illusion of water so realistic that the urge to dive into it was overwhelming. Even the towering ilala palms fringing it suggested that an ocean resort was at our fingertips, where suave waiters with muscled biceps would serve us frosty glasses of sparkling champagne. But Naomi and I could only dream of such things (and we did – often). The only water here bubbled up from underground springs, slightly alkaline but good enough for the animals to survive on.

By night we sat and chatted by the campfire, drinking Savannah ciders and cooking *wors*, spicy Namibian sausage, over the coals. Naomi was shorter than I am (which is really saying something), with long, flowing dark hair and a deep tan, but despite her girlie appearance and posh accent, she was right at home in the bush. She was a master fire-starter and chipped in with all the cooking and camp set-up. She was great company and full of fun.

On the first night at the old horse camp, I slept on my swag in the back of my ute, while Naomi rolled hers out on the ground by the back wheel. We learned that night that forgetting to put up our mozzie nets was a bad idea. Although it hadn't rained for over six months, raiding mosquitoes intent on fresh Aussie and Pommie blood devoured us. Dozens of tiny brown rose beetles invaded our beds, crawling around our scalps and tangling up in nests of long hair. Neither of us got much sleep.

On the second night we put up our mozzie nets, looking forward to a good night's sleep, only to find ourselves inundated by another creature – elephants. The herd of giant proboscideans pulled down trees and crashed around the fence that surrounded us, the sound of cracking branches and gurgling elephant stomachs preventing any sleep. It sounded as though there were dozens of them feeding on the periphery of our camp. The noise was overwhelming. We were totally surrounded, watching their enormous shadows as the herd lingered. Then the huge silhouette of a bull, higher than the tall wire fence, glided by. Our ears strained to hear the sounds of his slow footsteps, crunching on dried leaves and fallen branches.

'Maybe I should build the fire up,' I suggested to Naomi. 'Let them know we're here.'

In retrospect, I had totally underestimated the instinctive knowing elephants have when humans are nearby. We are so dependent on our eyesight that we sometimes forget that we also have hearing and smell to assess our surroundings. On this night, the darkness had unsettled us. All we could see were the daunting shadows of elephants moving through the night like ghosts, making deep gurgling and rumbling sounds, while branches cracked like gunshots and trees were pushed over. It sounded like a war zone. I built up the fire so it glowed like a large beacon.

'Shit!' Naomi exclaimed, causing my heart to miss a beat.

She was looking through her binoculars, which actually enhanced night vision, at the place where the crashing sounds were coming from.

'I think he's broken the fence! I can't really see properly, but... man, this is really freaky...'

With the cacophony of elephants around us, neither of us was brave enough to walk up too close to the fence to see if it was broken. My heart was beating like a drum, but I tried not to show Naomi, lest our fear build on each other's. If the fence was broken, the elephants and perhaps even lions and hyaenas would be able to come in to where we were sleeping out in the open. I was glad I wasn't there alone.

Feeling like a pair of useless, yellow-bellied girls, Naomi and I gathered up our sleeping gear and threw it inside Nad's tent. It contained many of his possessions, and we felt as though we were invading his private space, but it was the only option. All we could do was hope that we would never have to face the humiliation of a confession that we intruded because we were... ahem... *scared*...

The next morning, first thing, we checked the camp for elephant spoor. We felt pretty stupid when we realised that there were none of their dinner-plate sized footprints in the camp. However, all around the fence trees and branches had been pushed onto the metal barrier and it was broken in some places. The elephants had also broken a pipe and water was spurting out vigorously like a fountain. It did indeed look like a war zone, but the fence was still largely intact. To this day, we have never told Nad.

DRY WATER

Nothing is more indicative of the damage elephants can cause than their destructive path during the dry season when they are driven by the thirst and hunger that comes with living in an environment of vanishing resources. We were in awe. Etosha was showing us something of her harsh reality. But we hadn't seen or experienced anything of the real dry season yet. The most extreme was still to come – October, when everyone develops the dreaded condition known as October-itis.

SUICIDE SEASON

In Namibia pretty much everything was at its worst in October. People you could usually get along with – even if you didn't like them particularly – you couldn't *stand* in October. If you could put up with the plagues of tiny brown rose beetles that invaded your hair and formed crunchy bits in your rice in September, well, October pushed you past the point of tolerance. A niggle became an agony, a whisper a scream. Even breathing pissed you off. And it was contagious. Everyone in Etosha got October-itis at once. The only cure for this widespread condition was rain. Unfortunately, like most things in life, that was out of our control.

The interminable wait for the blessing of rain seemed to take forever. The mopanes stood naked, skeletons of their former luxuriant selves, casting no shade. Leaf loss in the hot, dry months saved a tree's precious moisture stores by reducing the surface area exposed to the harsh sun. Many plants in Etosha seemed to stunt their growth, perhaps because of water and nutrient limitations. The mammals, on the other hand, tended to be bigger than their counterparts elsewhere. In October, when the heat haze invaded my brain, I sometimes wondered whether I was more plant than animal. Was I stunting my growth or growing so rapidly that I just couldn't tell?

In this wildebeest-wilting weather, when a hypnotising blanket of heat rose up daily from the glaring white Etosha pan, there was so little browse and grass, my theory was that the animals became breatharians.

Dappled giraffe chewed on the sun-bleached bones of perished souls. Black-faced impala with ribs sticking out ate small calcrete rocks on the roads as if they were delicious sweets. Zebra carcasses with blood oozing from the orifices, victims of anthrax, seemed more conspicuous by the day as the season grew more extreme. Dry, hot winds swept violently across the landscape forming dust devils, carrying with them layers of desiccating white dust. When the dust fell, it seemed to choke the life out of everything, transforming the environment into something resembling a glaring snowfield. At this time, tension seethed beneath the surface of our day-to-day lives like a sinister amorphous monster. Like the plants and animals, we became subject to nature's every whim, seeming to teeter on the brink of insanity, waiting endlessly for the rain.

I was sitting at the staff computer emailing my mother one of the obligatory 'I'm alive' notes I sent every few days. My supervisor, Anne Goldizen, had just arrived from Australia to help me plan my project a little better. She'd managed to get the funding from the university and I needed all the help I could get on the scientific front. Anne knew as well as I did that I wasn't doing a PhD for the qualification: I just wanted to live in the African bush. And like so many zoologists, Anne had always wanted to see Africa.

Nigel rushed in and said that a tourist had just told him about some springbok fawns that had become trapped in the mud out at the waterhole. The Okaukeujo waterhole was and still is one of the most famous in Africa. At night I've seen up to ten black rhino drinking there, which is truly extraordinary for a species that is usually solitary as well as critically endangered. Herds of elephants streamed in from all different directions throughout the day; zebras hovered in the muddy water up to their midriffs, and springboks, less adventurous, cooled off in the shallows. The result of all of this frenetic animal activity was to turn the waterhole into a quagmire of mud and dung, especially in the shallows, so that animals were forced to walk in deeper towards the middle to find less muddy water to drink.

Nigel and I climbed over the rock and cement fence separating the waterhole from the camp, where tourists sat on bum-numbing wooden chairs or watched the animals from their rondavels. On the other side we squeezed through wires that provided the tourists a measure of protection from predators. We briefly checked around us for lions and then walked down to the waterhole, about thirty metres away. The zebras and springboks charged off upon our approach, splashing and stampeding their displeasure. Nigel was first to spot the three little heads, with sonar ears too big for their dainty faces, trapped in the mud. The soft, pale fur on their faces was splashed with mud, their eyelashes over large fawn eyes barely able to stay open.

When they saw us coming, they didn't even try to get away. Exhausted, their final energies totally depleted, the tiny creatures had given up and seemed resigned to their deaths. Even if they'd wanted to escape, the mud prevented any movement. It only sucked them under further. One fawn's nostrils were caked in mud and it was struggling to breathe. The fawns' big, brown eyes were so endearing and it was heart-breaking to see them trapped in a sea of choking mud.

Nigel whipped off his felt shoes and socks, hoisted up his beige shorts and waded in up to his thighs. Gently he handed each fawn, one by one, to me on the mud's edge. He wobbled precariously in the mire, almost being sucked under himself at times. As I held each fawn, I was amazed at how light they were, probably only a kilogram or so, and there was no fight left in any of them. Though they were all breathing and conscious, there was not a kick, not a bleat, not a sign of any will to live. I wondered how long they'd been stuck in the mud, fighting for their lives, before giving up. Their mothers were nowhere in sight as we plopped the three little fawns under a large leadwood tree near the waterhole, limp and lifeless.

They couldn't have been more than a week old. This was nature's typical harsh introduction to life in the African wild. The mud would soon dry and cake on their fur, so we tried our best to clean it off with our hands. We left them, exhausted, hoping that their mothers would return before a hungry jackal arrived on the scene.

As we scaled the fence, both of us covered in a mixture of mud and animal dung of every description, smelling like a fresh midden, we noticed all of the tourists clapping and cheering, Anne among them. There was a gooey dollop of mud on Nigel's Coke-bottle glasses, dripping onto his tanned cheek. The grinning, mud-speckled eccentric raised his arms above his head like a champion, acknowledging their cheers in good humour. He was only joking, but to me he has always been a champion. Later that afternoon we spotted a small, mud-coated springbok fawn on its feet and wandering around the waterhole among the adults. Nigel said the other two had probably been found by their mothers and taken back to the bush. I hoped so. It was an experience I will never forget, a reminder that in life there are always opportunities for small acts of heroism.

After a few days in Okaukuejo I decided to take Anne to the east of Etosha for a look at the black-faced impala population there, which meant camping overnight at the dreaded horse camp at Namutoni. It was Anne's first time in Africa and she seemed more youthful and carefree in this environment than she had in the stilted academic world of the university.

'I am having the time of my life, Tammie,' she announced one night, after I'd fed her a necessary Savannah cider. 'Thank you so much for organising this as your study site!'

Having Anne there was great for me too, because it brought into focus what I needed to do to complete my PhD. I was so euphoric to be living in Etosha that it was easy to forget I had to produce a doctorate at the end of it.

After a day of observing impala, we threw our bedrolls on a cement floor in one of the old horse stables which were now used for storage. There was a chance it could rain, so this seemed a better idea than camping out in the open. Plus I was trying to impress my supervisor with my great bravery and bush skills. Getting rained on at two in the morning probably wouldn't put her in a very good mood. And I certainly wasn't planning a repeat of my previous girlie escape into Nad's tent with Naomi. My reputation was at stake.

However, things didn't quite go to plan. I was sitting on a canvas camping chair writing in my journal and Anne was sitting on her mattress reading a book, both of us feeling quite content about life in general, when we heard a scuttling sound in the grass near the edge of the cement, about a metre from where I sat.

'What's that?' Anne asked in her usual calm manner with its crisp American accent.

'Oh, it's probably just a frog. There's been some rain up here,' I replied, unperturbed, and continued to write in my diary.

'Oh…' Anne replied. 'Lucky it's not a cane toad.'

We carried on with our respective activities. Then whatever it was shuffled again, louder this time. I glanced up from my diary and then looked at Anne.

In response to her questioning expression, I said, 'Sounds like a lizard… Don't worry, there's lots of harmless lizards around here.'

'I don't know, Tammie.' There was a glimmer of alarm in her usually calm demeanour. 'It could be a snake.'

I smiled and again told her not to worry, before looking down at the spot under the cement where the noise was coming from. Suddenly, on cue, from a hole under the floor, a two metre long western barred spitting cobra, commonly known as a zebra snake, slithered out. Finding two unwanted humans in the vicinity of its cavernous home, it raised its head up in typical S-shaped cobra fashion. The snake's body was almost as thick as my arm. Its black and white striped body was magnificent, but there was no doubt about its annoyance as its shiny, jet-black head, enlarged into a hood, hovered in the air, ready to strike or spit. Zebra snakes have a spitting distance of up to two metres and I was well within that range and not wearing sunglasses. I'd heard stories of how fast they could move when aggravated and knew that there was no antivenom for this species.

So much for impressing my supervisor with my bush knowledge.

'Shit!' I shouted and ran like hell towards the car without looking back.

Anne responded instantly and ran behind me. From the car, our hearts beating frantically, we watched the cobra lower its head and slither away into the undergrowth. There was no way we were sticking around to find out if it wanted to come back to its hole. Once again, I gathered up our bedrolls and sleeping gear and threw it all onto the floor in Nad's tent. Once again, I didn't tell him. A girl's gotta have her pride.

A couple of weeks later, with Anne on the plane back to Australia after a very productive project-planning trip, Naomi and I packed up the *bakkie* and made for western Etosha, about two hours drive from Okaukuejo. Heading into the west of the park always felt like an adventure. The only people we were likely to bump into were the anti-poaching team or the rangers. We wouldn't be running into any tourists asking us where the lions were, because this part of the park was for staff and researchers only. Tour companies couldn't drive through without a special permit.

As we drove along the wide, corrugated roads we plunged over towering dolomite mountains, past picturesque granite outcrops and through wide-open plains with patchy carpets of pastel yellow grass. The scenery here was much more spectacular than the flats of the tourist areas. There were fewer operational waterholes and less game. Nonetheless, the animals we saw fitted perfectly into the vastness of the landscape. Elegant, endangered Hartmann's mountain zebras and delicate tricoloured springboks scrambled across the rocky sides of large *kopjes*, along well-used game trails. It was obvious from all the large piles of dung on the road that elephants abounded here too.

We stopped at a waterhole with a large windmill spinning lazily in the mild breeze. A quartet of mud-caked warthogs fled as we drove into the parking lot, their tails pointed skyward. A family of ground squirrels, using their fluffy tails as parasols to shade their backs from the sun, squeaked alarm calls when we drove past, ducking swiftly into a hole in the ground. We left them to their own business and carried on towards the western camp, Otjovosandu.

It was about two in the afternoon and the October heat was choking. The hottest time of day, I'd discovered, was about three in the afternoon and the heat didn't even begin to dissipate until after dark. The heat, combined with the dryness of the air, turned our hair into ragged straw mops. Our lips cracked and required constant moisturising, our heels cracked and our tanned legs were dried out and covered in a fine white powder. Naomi and I were the only eligible females in the vicinity, but we sure didn't feel like we were. Etosha was turning us into wild bush witches of the west!

We'd been driving for an hour or so, with the glare rising up off the white gravel road and the heat haze becoming mesmerising, when for no apparent reason the car spluttered and broke down. I let it roll to a halt. I was too hot to care.

'Uh-oh,' I muttered, raising an eyebrow at Naomi.

I didn't have the energy to say anything else and nor did she.

I checked the temperature, but the gauge indicated that it wasn't much higher than usual. I wiped the sweat from my brow with my light shirt. We were at least an hour from Otjovosandu camp. Bugger. I tried to look knowledgeable by opening the bonnet of the car. The engine was hot, but there was nothing unusual about that. After all, *everything* was hot. It was suicide season.

I left the bonnet up and scanned my surroundings. We were less than a kilometre from Okawao waterhole, where there was a staff campsite enclosed by a fence, but there was no way we were walking that far on foot. Western Etosha's lions were notoriously aggressive towards humans. Shayne and Birgit had told me about lions chasing vehicles, jumping onto bonnets and crashing through windscreens in this area of Etosha.

Low mopane shrubland without leaves cast no shade upon the red sand dotted with calcrete rocks and shrivelled dead leaves. The sky above us was white, not blue, from months of dust devils and whirlwinds. We'd just been driving through an area where recent bushfires had reduced the remaining vegetation to blackened stumps and grey ash. It felt like hell on earth.

Naomi was peering into the engine, but I knew that neither of us had a clue about mechanical things. We were grounded. It was after five, which meant that the warden and rangers at Otjovosandu had knocked off for the day. I tried to radio them but there was no answer. I was too far from Okaukuejo to get through to Shayne on his radio. We swore horrifically like the mad bush chicks we were becoming. Then we did the only thing that one can in situations like this. We laughed hysterically and decided to make a plan.

Naomi threw some firewood on the gravel road and managed to start a fire. We drank lukewarm Savannah ciders, cooked some fatty *wors* and sweet potatoes in foil over the coals and ate the hot sausage with bread as the sun was going down. The fire gave us small comfort as we heard elephants trumpeting at the waterhole nearby and then, the sound we'd been dreading, the guttural *humph* of lions. The big cats sounded close, perhaps even at Okawao, but we couldn't tell because lion roars can be heard from up to ten kilometres away. When it became too dark to see, with our nerves on edge, we got back into the front of the single-cab ute and settled in for the night. It was sweltering, but neither of us wanted to risk being pulled out of the cab by a hyaena or a lion, so we wound our windows most of the way up and put up with the heat.

'Matson, you're such a wimp... I mean, *really*, you black-faced impala researchers... too scared to sleep out in the open!' Naomi jibed as lions roared in the background.

'Well frankly, Pommie Girl, if you're so brave, why don't you sleep in the back of the *bakkie* so I can stretch out here in the front? Come on, you worked on leopards, you're meant to be hard core...'

Trying to make light of the situation, we chatted until we fell asleep, our legs cramped awkwardly in the small seats. I was woken up constantly throughout the night by the loud roaring of lions, but luckily they didn't seem to realise we were there, or perhaps they just weren't interested.

The next morning we hung a green canopy off the side of the car and pinned it to the ground with rocks to give us some shade. The heat grew more and more unbearable as we waited for someone to drive by. A parks vehicle was bound to come past eventually. Luckily it was a weekday, so we wouldn't have to sit here for several days – hopefully. By midday we were about to spontaneously combust. It felt as though we were being cooked in the oven of hell. Annoying mopane flies buzzed around us, trying to invade our eyes and nostrils. I really wasn't in the mood. This wasn't fun any more.

Just when we thought it would never end, knowing that the worst heat of the day was still to come, we saw an apparition in the distance. I thought I was hallucinating. A massive bus full of German tourists materialised out of the haze. The driver, a friendly looking man with a grey beard, stopped the bus when he saw us, two silly, dehydrated girls stranded beside the remnants of a campfire and a *bakkie* with its bonnet still up. Naomi and I were euphoric.

The guide introduced himself as Klaus in a thick German Namibian accent and asked us what the problem was. I explained that we really didn't know, but that since yesterday we'd been broken down there.

'Haf you tried to start it since yesterday?' he asked.

'Um…no… I think I may have flooded it from trying to get it going yesterday,' I replied, hating the fact that I had no mechanical knowledge.

Klaus settled into the driver's seat and put her into neutral, while his busload of tourists watched in fascination. They were snapping away on their cameras. He turned the key and, to my utter amazement, she started! Naomi and I looked at each other with expressions that said: this is the most humiliating moment of our lives.

'Ja, I don't zink you haf problem. In dis hot weather, ze car is getting very hot and you must just let it sit for some time with ze engine off, ja? Zen it is starting okay,' Klaus explained matter-of-factly.

I nodded with embarrassment.

'You see, if you were starting it yesterday after letting it rest a little while, then it would haf started,' he went on, further amplifying my humiliation.

Life in Etosha was all about adaptation. To stay fit, Naomi and I began to jog around the fenced perimeter of Okaukuejo camp each afternoon, a distance of about three kilometres. Running in Etosha was hard during the dry season because we inhaled as much dust as we did oxygen and it wasn't cool enough to run until almost eight at night. By then it was dark, there were no streetlights and you ran the risk of running into a rabid jackal. Sometimes, on our jogs, dust devils tore past us and at other times we had to run right through them. Our clothes stank of sweat and dust. One day, we realised we were running beside a herd of twenty-six elephants on the other side of the fence. Elephants of all sizes and shapes were showering themselves with dust. We stopped running to watch them. As a blood-red sun sunk into a sunset that seemed to turn the sky to fire, their giant silhouettes in the dust became a blurry haze of elephants, swirling dust and fiery light. Their fuzzy forms glided ethereal in the orange and golden radiance of sunset. I thought to myself, this is why I love it here.

As the hot, dry season wore on, Naomi and I both had to admit we were developing a distinct admiration for the black-faced impala. With their food sources drying up, the daily quest to find enough vegetation to survive took its toll. By late October, the female black-faced impalas were heavily pregnant, so on top of having to survive the predators at night, by day they were struggling to find enough food to feed two in an environment almost devoid of sustenance. Lethargy hung over us all, people and animals alike, threatening to swallow us whole with despair at nature's cruelty. As we monitored the behaviour of the impalas each day, we felt their enervation and shared their longing for the rains to come and ease their struggles.

Life thrives on so little in desert environments. A small shower of rain is all it takes to trigger a blooming field of flowers. Like the animals and plants in this arid environment, I was becoming a person who needed less to survive. In fact, I was thriving on the quintessential simplicity of it all. Trendy clothes didn't matter. Expensive food and drinks seemed a waste of money. My small, dusty caravan seemed a perfectly adequate home. I owned nothing of material value. I remember thinking that I had never felt so completely content in my life.

Being out in the bush each day, Naomi and I saw first-hand how everything was connected to each other and to the evolving seasons. Over several weeks we monitored the progress of three tiny jackal pups we discovered inhabiting a termite mound beside the road on our route to Olifantsbad waterhole. Each day we stopped to watch them rolling around and playing with each other, their fur still fluffy and soft-looking. Sometimes their parents would be with them, but often they left the pups alone while they went away to hunt. We watched their mother returning to regurgitate food into their begging, wide-open mouths. Next year these jackal pups, if they survived, would be hunting the lambs of the impalas that were due to arrive at the beginning of January. They too had to make a living as part of the ecosystem.

Another time we noticed a spotted hyaena loitering beside the road. I stopped the car about thirty metres away from it, expecting it to take off. But the shaggy, inquisitive creature, with its hunched back, drooling, deadly mouth and crunching, canine teeth was curious. It walked up onto the road, wandered around the car, sniffed the tyres and then proceeded to poke its head in the passenger-seat window where Naomi sat with her eyes widening by the second. To this day I have never seen another hyaena show so much curiosity towards humans. He was actually checking *us* out.

Magical moments seemed to surround me. I was walking back to the research camp from a sundowner at the Okaukuejo waterhole one evening when I heard a child singing sweetly. I was walking past the staff housing, a conglomeration of cream-coloured old caravans. In the dusty yard of one, I noticed two little black legs sticking out of a rusted wheelbarrow.

'Jeeeeesus loves me... Jeeeeesus loves me...' the little boy was singing in a gentle, high-pitched voice, kicking his legs around merrily in the wheelbarrow, with not a care in the world.

I was in Etosha to learn about black-faced impala, but I was discovering so much about everything else. I felt ready to absorb anything and everything that Africa wanted to show me. I'd long since stopped being a practising Catholic, to my parents' disappointment, but I felt myself becoming a more spiritual being the longer I spent in the African bush. Although I didn't want anything to do with institutional religion, spirituality permeated my life because I was ensconced in nature. All the answers were there.

After four months, to my delight, I was upgraded to a newer, bigger caravan in the research camp after a pair of German termite researchers – the Germites, we called them – moved out. My new caravan was airconditioned, and as far as Naomi and I were concerned, we'd never seen such luxury. It had been donated to the camp by American researchers from the San Diego Zoo over a decade before. The quaint mobile home had an operational fridge in a kitchenette with a small table where I set up my laptop. A lounge pulled out into a comfortable bed for Naomi and I had the absolute indulgence of a double bed. A tin roof over the top of the caravan extended over a cement area to one side, where we set up a couple of canvas camping chairs and a metal table. The interior was straight out of the 1970s, a combination of wooden walls and shelves, cream and brown wool on the couches and gold light fittings. We hung up sarongs from Byron Bay, photos of our friends and families and lit a stick of incense. Now Etosha really felt like home. But we couldn't get too settled in because I still had a lot of work to do out on the farms.

If we wanted to build up the population of black-faced impala, we would need to translocate small herds to new areas, and to do that we needed to know what had affected the success of previous translocations. That's where the farms came in. There were about twenty of them with translocated black-faced impala, located all over the country, from the mountainous area west of Etosha to the sandy south-east of the Kalahari Desert. I would have to visit them all, interview the owners and look at their farms and black-faced impala populations in order to gain a full understanding of how to maximise the success of future translocations. It was a daunting prospect, not least because most of the farmers spoke only a little English. Many of them lived in remote areas, some of them largely cut off from the outside world, and I never knew what I was getting myself into.

It was a long drive from Etosha to the Kalahari and, as usual, we took a few wrong turns on winding dirt roads, navigating our way from sketchy phone instructions in haphazard English, before we found the first farm. The psychedelic red sands of the Kalahari

Desert were dotted with stout, black-thorn acacia bushes and tall, arthritic, flat-topped camel-thorn trees.

It was in this colourful, contrasting desert environment that we found the gate to the Martin family's farm. As we drove in, a semi-tame giraffe towered over us, leaning in to take a good look at us. When Naomi got out of the car to open the gate, it leant right over to her as if to kiss the top of her head, so close she could smell its musty breath. We carried on up the red-dirt track to find the family's quaint farmhouse. Like an apparition in the midday sun, white swans floated elegantly upon a circular, man-made canal surrounded by luminescent green grass. There was a table and a set of chairs on a grassy island in the middle of the canal, like something straight out of the colonial era. Naomi and I blinked, thinking we were seeing things. Green grass and white swans were two things you didn't see much of in Namibia.

With the same warm generosity I'd experienced at farms throughout Namibia, Mrs Martin invited us to join them for lunch. Her husband, a round, rosy-cheeked Afrikaner with a kind face, said he would show us his black-faced impalas after we'd eaten. Naomi and I accepted their invitation and sat around the circular dining table with them and their four blue-eyed, bright blond children ranging in age from about eight to sixteen. The children were gawking at us as if they didn't see other white people very often. The family seemed to come from another era, tucked away in their own private world of sand dunes and white swans.

An African woman, their cook, brought out a magnificent spread of mutton roast, gem squash and other homegrown vegetables. It was a feast and I wondered whether they were eating like this for our benefit or if this was how they ate every day.

I wondered why all four children were at home, and asked the older girl if they were on school holidays.

Shyly she responded, 'No. We have school at home. It is the American Christian Education system.'

I asked her how often they visited the nearest town.

'Maybe once a month,' she answered, lowering her eyes subserviently.

It was a sheltered life, but they all seemed so content and peaceful in their devout, simple farming life.

Naomi commented later that she'd been tempted to swear and exclaim, 'Hell, you lot don't get out to party much, do ya?'

During all of my visits I was astounded at the generosity and warmth of Namibian farm folk. To them I must have seemed like a bit of a weirdo. I lost track of the number of farmers who asked me why I, a young girl from Australia, was wandering around Namibia looking at black-faced impala. Many of them didn't even know how unique and endangered the subspecies was, so my farm visits became as much an opportunity to educate the public about the subspecies as it was to educate myself.

As it turned out, the farm visits resulted in some of the most useful information to come out of my research. If it wasn't for all of those wonderfully eccentric, open-hearted folk throughout the country, we'd still be wondering how to write the black-faced impala management plan. The history of translocations of black-faced impalas to farms in the last thirty years showed that to ensure a successful population, a founder population of at least fifteen animals was needed. Fewer than that and almost certainly they would be wiped out by cheetahs and other predators. What began as a complex situation turned out to have a relatively simple solution. It was a game of numbers.

ONE OF THE BOYS

My stomach was churning and I wasn't at all certain we were going to come out of this alive. When Nad Brain, he of the infamous tent, had invited me to join him to look for a radio-collared cheetah in the ministry's light plane, I'd jumped at the opportunity. As a kid, I'd flown in four and six seater single-engine Cessnas with my father, who flew to remote outback properties in his job as a property valuer. So I didn't think for a moment that my stomach would disagree with the flight. But in Etosha, with its gripping heat and twirling dust devils, not to mention Nad's notorious flying, the inside of the plane quickly became claustrophobic and stifling. Nausea overcame me as Nad flew low over the carpet of trees and then swung the plane to one side and dived down to take a closer look at something. I breathed deeply and looked out the window, focusing on the horizon, willing myself not to throw up.

We were flying low enough to spot herds of antelope moving along the intricate matrices of game trails carved through the mopane shrubland and joining up at the waterholes. Game concentrated at some waterholes in the hundreds – oryx, springbok, ostrich, zebra and warthog all lined up to take their turn at these oases. We flew low over a herd of elephant tearing apart mopane shrubs with apparent ease as they meandered through the bush. Tiny baby elephants hid beneath their mothers' gargantuan bellies. The scene was mesmerising and took my breath away, despite my stomach's objections.

When we landed on terra firma, Werner, another scientist in the institute and Nad jokingly commented that I looked a bit white for a tough Aussie. I was wobbly for the rest of the day, but determined not to show it. I had a reputation to uphold.

After a year in Etosha I'd become part of the furniture. I was beginning to feel asexual, like one of those self-sufficient, solitary desert plants that needs only an appendage to be taken off and planted in order to reproduce. I was just one of the boys, Nigel said, which made me feel both flattered and completely androgynous. I'd always been a tomboy, but this was taking it to the extreme. After Naomi had left for England, I'd immersed myself in the social scene of the Okaukuejo locals. It seemed I'd finally been accepted as one of them, because most researchers came and went after a few months but I was still there. Even the boys' club in the tearoom was becoming less intimidating.

The turning point came the day I decided to do something completely out of character. I baked a cake.

I'd baked it in the dubious gas oven in the research camp, which had become a cosy home to a family of mice. The kitchen housed a haphazard array of old saucepans without handles, chipped mugs and plastic plates with deep scratches. It had the look and feel of an ill-kept communal kitchen and at night it was the hub of activity. The pungent stench of burning mouse faeces permeated the kitchen as the cake cooked. I was sure it was going to be a holy disaster. To my surprise, however, the chocolate cake didn't reek of rodent dung at all. In fact, it was positively edible! I topped the brown sponge off with a layer of chocolate icing and declared it ready to face the tearoom.

I floated in with my creation, ignoring the blatant stares, and placed it proudly upon the table next to the teapot. No one uttered a word. It was as though there had just been an earthquake and everyone was too scared to say something in case it happened again.

'Morning, guys!' I announced, smiling winningly. 'Anyone for chocolate cake?'

Wilfred, the research technician, was staring at me as though I was some strange circus act, but there was a flicker of amusement in his eyes. The other scientists looked vaguely perturbed. Suddenly Coen stood up and walked over. Without looking at me, he leaned over the cake until his nose was almost touching it, closed his eyes and inhaled deeply. A wide grin spread over his face.

'Hmm…' he said, rubbing his bulging belly. 'Chocolate…my favourite.'

The awkward silence was shattered as all the men began to bellow with laughter. Coen had leaned over a bit too far and a dollop of wet chocolate icing was stuck on his ferrety moustache.

'I'm saving it for later,' Coen announced, not bothering to wipe it off, then indicated the cake: 'May I?'

Chocolate cake was served and the conversation slipped to English and became animated. I couldn't believe it.

'We didn't know you could cook!' Nigel exclaimed, wolfing down his second slab. 'This is great!'

'Hmmph. Very nice. Yes, good cake,' announced Wilfred seriously, assessing it like a scientific report while crumbs sprinkled all down the front of his uniform.

'This is great, Tam,' said Shayne. 'You're not trying to poison us, are you?'

If only I'd known. Chocolate cake. That was the answer.

For a change, the half-hour tea break flew by very quickly until everyone shifted from their seats and headed back to work. After three pieces of cake, Coen was the last to leave. He stopped as he reached the door and turned back.

'Thank you, Tammie,' he pronounced, smiling shyly. 'I like your cake very much.'

Then he turned his back and departed in his regular authoritative manner, licking icing off his fingers as he went.

I fought the urge to jump up on the table and scream 'Yes! Yes! Yes!' at the top of my lungs.

After work each day, when it cooled down enough, Shayne, Birgit and I established what would become the closest thing to a social sport that Okaukuejo could muster. We called it Okaukeujo hockey, a game played with two teams wielding ice-hockey sticks improvised from broomsticks on a tennis court with two goals made of netting. The rules were there were no rules. You could hit the hockey ball as hard as you liked and it didn't matter if it hit anyone's feet as it does in normal hockey. Bruising was compulsory. Exhaustion was inevitable. It was vicious, sweating stuff that left us all puffed and battered and ready for the obligatory Tafel lager that followed. Soon we had teams of over five on each side and Okaukuejo hockey was even attracting players from the neighbouring farm. It kept us all fit and invoked a sense of community spirit that seemed to have fallen away.

In the institute Johan le Roux, the resident Geographical Information Systems expert, offered me the use of a computer and his lab to back up my data and start working on producing maps of impala distributions in the park. With his serious exterior, I never knew quite where I stood with Johan. This was a man I'd seen make women researchers cry. But I soon learned that he meant his criticism to be constructive and I would never have been able to understand the way computer mapping systems worked without his long hours spent teaching me the ropes with infinite patience. A tough exterior seemed to go with the territory in Okaukuejo, but I was learning that none of the men was really like that underneath. It was a harsh environment, isolated and hot as hell, and people became hard (at least, on the surface) in order to deal with it.

Werner Kilian was a prime example. Unmarried and in his forties, Werner was politely professional on the surface but marshmallow underneath and a sucker for good red wine. He invited me to join him and Rainer and Stu, two game managers from Ongava, the farm next door, to make up his team for the black-rhino monitoring in western Etosha.

To my joy and my supervisors' worry, the great thing about living in Etosha was that there was so much more going on than my own research. I'd helped immobilise a lioness to take its collar off, only to have her wake up and snarl halfway through the process, causing three macho men to momentarily lose all their dignity and run for the hills. I'd joined Shayne, Nigel and an Austrian researcher to remove a collar from a spotted hyaena after the men had dragged a gutted springbok around to coax her to the bait. And now I had received an opportunity to help research the black rhino of Etosha as part of routine park monitoring. Up until then I'd been jealous of my sister, Kek, a vet student, who'd helped immobilise a rhino in Etosha when she visited. Now it was my chance to get up-close and personal with some rhino and I was ecstatic.

Determining the population size of one of the world's most endangered – and dangerous – mammals... It sounded easy enough. But that was before I got my hands dirty.

Hurdle number one: black rhino drink mostly at night, necessitating that counts be undertaken between sunset and 1 a.m.

That means you are out there when the predators are hunting – in the dark, when all self-preserving humans are tucked up snugly in bed. This wouldn't be a problem if you were safely inside a vehicle with all the windows wound up. However... Hurdle number two: you have to get close enough to the rhino to take a mug shot. Most of the rhinos in the park were identifiable by distinctive ear notches, so in theory all that was needed was to take a photo of the rhino when it came to drink and to record its sex, age and other noteworthy features. This meant you had to walk *away* from your car, down to the waterhole and photograph the rhino at a distance of a mere twenty metres from the rhino's left eyeball. To put it mildly, it was not work for the lily-livered.

Usually, at each waterhole, teams worked in groups of two per car. One person walked down to the waterhole and took the photograph, while the person in the car recorded the rhino's characteristics and acted as back-up in case the animal charged. Theoretically, the process was simple, 'a piece of cake' the old boys said. In practice, rhino monitoring was one of the most dangerous and exciting jobs you could hope for.

On the first night we were sitting at Jakkalswater, an Afrikaans name for a waterhole known as a drinking place for jackals. Werner's beige Land Cruiser and Rainer's white Hilux were parked about fifty metres away overlooking the waterhole, where earlier we had set out piles of white calcrete rocks as markers. Each marker identified a twenty metre mark from a possible drinking point of rhino, so that we would know the correct distance to take photos. This was based on the assumption that human judgements of distance are poor at night, and worsen as the night draws on. The distance had to be exact so that the camera, pre-set to focus at twenty metres, took sharp, clear photos.

When we'd arrived at Jakkalswater in the late afternoon, lolly pink clouds had washed across the massive sky. Beneath its rose-coloured glow, reflecting in the water, eight elephant bulls were relaxing. They had laid their serpentine trunks over each other's rough backs and seemed to be enjoying the light breath of the winter wind as the final rays of daylight faded. We waited for over an hour for them to drink and leave so that we could lay out the markers before it got too dark. But as elephant bulls tend to do, they loitered lazily at the waterhole, not drinking, just enjoying being in the presence of each other at the gentlemen's bar of the bush.

As darkness began to descend, Werner decided we'd have to force them to leave. I held on tightly in the passenger seat of the Cruiser as Werner revved the engine and charged at the bulls a couple of times. They let us know they weren't happy with our bullying with plenty of ear flapping and trumpeting, but after a couple of charges, they grew tired of us and sauntered off. We laid out our markers hurriedly and rugged up for the long night ahead.

I pulled my striped woollen balaclava over my ears, which were bearing the brunt of the wicked evening chill. Despite the fact that our windows were wound most of the way up, our bodies fell into bouts of metabolic torpor when there was no action at the waterhole. The gentle hush lulled us into a quiet sense of camaraderie, intercepted regularly by Werner's deep grunts of boredom when there was nothing happening.

'You know, I think you're scaring all the rhino away with that balaclava pulled over your grumpy face,' Werner jibed, as we both shivered in the cab of the car. It was so cold even my teeth ached.

The four of us buoyed each other up with steaming cups of cheap instant coffee. Stu, a rugged, sun-blond surfie who was the assistant manager of Ongava Game Reserve, laced his with Old Brown Sherry for the extra heating effect. The hot liquid helped to defrost our innards and the caffeine kept us alert. None of us could afford to be anything less than absolutely vigilant when a rhino came to drink. Luckily, with Stu and Rainer there too, Werner had three sets of eyes looking out for him.

This was particularly useful early on in the night, when we were paid an up close and personal visit by a pride of lions. Rainer, a hulk of a German, had wrapped himself in a doona and was standing beside the Land Cruiser scanning the area for rhino. Stu, who was about my age and level of maturity, and I jumped up and down on the spot like Maasai warriors to try to stay warm. The two older blokes looked at us like the cold air had invaded our brains and caused temporary dementia.

A little while later Stu said something muffled that sounded like, 'Rhino.'

He was walking back to the car in a hurry, alarmed that none of us was getting inside because what he'd actually said was, 'Lions.'

The second time he said it, we all caught on very fast. The silhouettes of two young males and two young lionesses in the dim light of the full moon became visible as they crept up behind Stu's power-walking form.

'Get in the car!' Werner commanded and no one argued.

We all dived into our respective vehicles without a thought for what we'd left outside. Rainer's camera was still perched on a tripod beside our car. It was said that lions in western Etosha were not afraid of humans, and I wondered if these would live up to their reputation.

The cats sauntered up to the Cruiser in which Werner and I sat shivering, in my case from fear now as well as cold. I looked out the back of the cab window to see them sniffing around the vehicle as if it was some strange novelty for them, inhaling the plethora of human smells on the tyres. I sat very still and neither Werner nor I spoke. One of the lionesses began smelling Rainer's tripod. The look on the cat's face was one of pure fascination.

Then she swiped it with a feline paw to see how it would react, as if she thought it might be alive and possibly edible. They were all young lions and very curious. Luckily, the tripod stayed put and the lioness grew bored of it within a few minutes. Instead, the four lions slumped down to rest behind our car, just metres from where we sat. We couldn't believe it. They felt so comfortable next to the cars that they decided to watch the waterhole from there, perhaps even using the vehicles as cover from which to hunt. Werner turned to me and smiled. We both knew that this was going to make our rhino monitoring a little interesting.

'You still wanna photograph the rhino?' I whispered.

It was incredible to see lions so close up, but they were going to make for an unproductive night of research. It was unlikely the rhino would come in with the predators there, and even more unlikely that any of us were going to get out of the car to take the necessary identification photographs if one did come in. For about twenty minutes the lions lay beside the cars, simultaneously magnificent and menacing. Then for some reason they left. Perhaps they sensed the frustrated humans in the cars or maybe they left in search of better hunting grounds. We were relieved, but it was quite a while before I felt brave enough to venture outside the car again.

At ten, the night was still. Not even the cry of a black-backed jackal interrupted the hush. I raised my binoculars and scanned the bush around the waterhole, looking for changes in the moonlit landscape, small flickers of movement in the darkness. The full moon reflected in the silver water so brightly it was possible to pick out the individual shapes of leaves on trees, swaying slightly in the light breeze.

This was when rhino monitoring became dangerous, when tiredness set in as a result of inactivity.

Suddenly, through my binoculars, I noticed a dark grey blur emerging from the woodland. Rotund and heavy-footed, the blur morphed into a distinctive shape in the clearing, the crackling sound of dislodged rocks heralding each step.

'Rhino, Werner,' I announced, keeping my binoculars focused on the animal's position.

It had stopped now, surveying its surrounds for any sign of danger before it rumbled the last thirty metres to the water's edge.

Werner yawned and muttered, 'Well it's about time. I was just about to nod off.'

He opened the Land Cruiser door slowly to avoid its hinges creaking. Then he hung the strap of his camera around his neck, heavy and awkward with its bulky flash, and began the slow descent towards the waterhole. The rhino still had not moved position but had turned to face away from the water, broadside to me. I could see now, as Werner stepped tentatively closer to the marker, that it was a bull. I squatted on the ground beside the car, balancing awkwardly with one hand leaning on a hard rock, watching the animal. Two horns, one near the front of his head and a larger one further back, adorned a beak-like, almost reptilian head, forming a majestic silhouette against the moonlit backdrop. A mature bull, his horns had been worn down by years of wear and rubbing on stones and trees, forming a stump where once would have been a sharp, lethal point.

I couldn't see any ear notches, but I knew that Stu and Rainer would also be looking for them in their car, hopefully while also scanning for elephants and lions. Werner looked confident, but it was impossible not to feel some adrenaline.

I had done the walk to the water's edge to take a rhino's mug shot only once, with Shayne by my side, when I'd joined him and Birgit on the rhino-monitoring training session. I recalled how a pang of fear had shot up through my stomach into my chest, tightening my throat. I'd thought about what else might be lurking out there as I squatted at the waterhole, twenty metres from a black rhino. I felt like a lowly human, more vulnerable than the lowest animal on the food chain. I had been struck by the thought that even an impala, common prey for most predators in much of Africa, could run faster and jump higher than I could and, what's more, had horns to defend itself. Our safe man-made world makes us largely invincible to predators in towns and cities, but here in the natural world we were prey. These thoughts had made me lose my focus, blurring my vision, weakening my resolve.

I had walked slowly with Shayne at my side, stepping carefully to avoid knocking rocks. I'd crouched down, leaned my hip against the pile of rocks and raised the camera. Click, FLASH! The rhino had snorted loudly, rocks scattered, creating a sudden unsettling stir. He'd snorted again. I'd forced myself to remain rooted to the ground, the blood surging through my veins. I'd taken another photograph, blinding the rhino with the flash. That was enough, Shayne had indicated with a cutting gesture of his hand. We'd retreated then, walking backwards slowly, the rhino still huffing occasionally on the other side of the waterhole. After a few minutes, he'd begun to drink again, as if the disturbance had never occurred. Adrenaline had surged through me then, but I'd had Shayne with me, armed with a handgun, and Birgit's eagle eyes covering our backs in the car.

Now, with Werner flashing the rhino at the waterhole, he was depending on me and the other guys for his safety. No one could afford to make a mistake in this nocturnal environment with threats you couldn't see.

'Nice one, Werner,' I said when he got back to the car.

'Did you get the notches?' he asked, grinning.

He offloaded the heavy camera gear onto the passenger seat and leaned over to look at the data sheet.

'Number seventy-two. Three notches on the right ear and one on the left,' Stu interjected. 'He's an old boy by the look of those horns.'

Werner nodded. He was enjoying himself.

As Stu poured us all a warming Old Brown Sherry I contemplated what other twenty-three year old females were doing in the 'real' world. It was Friday night. Elsewhere girls my age were dancing the night away in clubs, being chatted up in pubs, throwing back cocktails, watching the latest chick flicks at the movies, going out to dinner at cool restaurants, grooving with funky bands... Here I was, freezing to death in the middle of the African bush, accompanied by a middle-aged bachelor taking photographs of rhinos, and hoping that a lion wasn't going to eat me... Life may have become a little unusual since I'd made my home in Etosha, but frankly I wouldn't have wanted to be anywhere else in the world.

Werner mumbled something under his breath, scratching the rough black stubble on his chin. It was late and sleep deprivation was setting in. Yawning, he yanked off his woollen beanie and threw it on the seat, to prevent any reduction in hearing. You needed all your senses to be on guard – particularly hearing – because even with the best night vision, the likelihood of you actually seeing a stalking lion before it was on you was minimal.

Werner's stocky form was halfway to the waterhole before I'd even registered he was gone, stalking the rhino who had just arrived. Stu was outside my window, leaning his arm nonchalantly on my door, looking through his binoculars. Unlike me, he seemed to be wide awake and on guard. I scanned the perimeter of the clearing with my binoculars. Nothing threatening. Other than the lions' visit the watch had been fairly uneventful and I didn't expect things to change now. We'd had five rhinos come in at different times, mostly in the early part of the evening. Three flashes in close succession and Werner was on his way back up to the car.

The rhino had been unperturbed by the flashing camera and the human presence. She'd stopped drinking, snorted and looked up, startled a little but not sufficiently to charge or head for the bush. I noticed that her first horn was long and skinny, stretching up and back to a sharp point like a large claw.

'Yesss!' Werner exclaimed when he returned to the car, 'Did you check the size of her horn? Never seen anything like it!'

I watched her in the moonlight. A representative of a critically endangered species, of which Namibia has the largest free-roaming population in the world. I was looking at an animal worth hundreds of thousands of dollars, a primitive-looking, half-blind creature that had been cursed by its facial appendages, unlikely phallic symbols perched on thick, wrinkled skin. A single rhino horn fetched a fortune on the black market, thanks to the demand for it as an aphrodisiac and for the handles of daggers. That was why protecting the Etosha population was so crucial. These rhino represented a large chunk of the global remains of the population. Furthermore, the rhino in Etosha were a unique subspecies, adapted to and only found in the arid environments of Namibia.

It was one o'clock in the morning and I was bleary-eyed with exhaustion. We were miles from the nearest campsite and in the most remote part of Etosha, so I pitched my tent beside the other guys, who slept out in the open on bedrolls, beside a water tank surrounded by a concrete wall. We all knew that lions and hyaenas could easily jump the concrete barrier, but we were too cold and tired to care. After a nightcap of Werner's port, even the full moon, golden and luminous, couldn't keep us awake.

After three nights of rhino monitoring, we were all worn out and bedraggled. We drove back to Okaukuejo desperate for a shower and a good night's rest. But as I was about to head back to my caravan Stu asked me if I would help them at Ongava that night for their rhino monitoring. The moon was still full enough to see by and Ongava had a number of black and white rhinos they monitored regularly. After a warm shower, the idea didn't seem so daunting and I drove over to Ongava to join the guys. Stu paired me up with Ongava's new maintenance manager and mechanic, Tristan, otherwise known as Bushman. I knew his older brother, they came from one of the most famous conservation families in Namibia, the Coopers. The two boys and their younger sister, Victoria, had been raised in the Waterberg Plateau Park in central Namibia, which their parents had turned into a thriving wildlife reserve in the eighteen years they'd lived there.

Tristan was a year younger than me, a smoker, with a shaved head and earrings and I recall thinking that he looked more like a Belgian tourist than a guy who'd spent his whole life in the African bush.

It was three in the morning and I'd been dozing when the only rhino we'd see all night came in. Tristan shook my shoulder gently and told me it was time to take the shot. Walking down to a black rhino in the dark tends to wake you up in a hurry. As the cold air hit my face and I spotted the shadow of the rhino, I went from being in a deep sleep to total adrenalised wakefulness in about two seconds. Three minutes later we were crouched on the opposite side of the waterhole to the rhino at our twenty-metre marker. The rhino snorted, scaring the shit out of us. I took the shot, thinking that he was either going to charge us or run the other way. The camera flashed, blinding the rhino, and he snorted again and shuffled indignantly. Tristan laid his large, calloused hand on my shoulder to steady me, knowing that I was seconds away from running like hell. The rhino snorted again and we both jumped out of our skins. Then he turned on his haunches and disappeared into the night, leaving Tristan and me looking at each other and wondering what we'd been so worried about.

SURVIVAL OF THE FITTEST

'Shit!' Tristan said vehemently. 'She must have deserted this one too.'

We were sitting by Ongava waterhole, as we had for the past twenty-four hours, as part of the reserve's annual seventy-two hour waterhole count. Every animal that came to all eight operational waterholes had to be counted and sexed in order to provide estimates of their population sizes. Rainer and Stu had been short of counters, so I'd volunteered to help and had been quickly paired up with Tristan. I had a feeling he'd organised that we would be partners on the game count. I'd sensed that he was interested in more than just a friendship with me; I wasn't sure how I felt. I didn't think he was my type, but ever since I'd moved to Etosha I wasn't sure what my type was any more. I was beginning to think I should let my guard down and just go with the flow. And Tristan certainly had potential!

During the day we'd counted literally hundreds of animals. Herds of zebras, oryx, black-faced impalas, eland, kudu, warthogs and waterbuck swarmed around the waterhole in an almost constant surge of life. It was almost impossible to keep count of all of them. While Tristan would be counting a flurry of oryx at the water, I'd be counting and sexing an incoming herd of black-faced impalas, while both of us struggled to monitor which zebras were newcomers and which had been there for a while, so that we didn't count them twice. After many hectic hours the animals became a blur, a whinnying, barking, coughing cacophony of hooves and hide. When the sun set, the concentrations of prey animals subsided and gave way to the nocturnal drinkers, the black rhinos, lions, cheetahs and, just occasionally, leopards. For a brief while, though, the waterhole was silent.

In this transition from day to night, a small lion cub came to the water's edge. The young female was alone. She drank as though she was dying of thirst.

'This is the sister of the cub that Rainer shot a few weeks ago,' Tristan explained, watching her through the scope. 'The one the lions broke the back of, remember?'

'How could I forget?' I replied, remembering the day clearly.

To put the cub out of its misery, Rainer had shot the mangled, paralysed animal, its back broken by an attack by an adult lion. Its mother, perhaps realising that her cub was doomed, had deserted it.

'Now the female must have ditched this cub too. Shame. She lost the lot this year,' Tristan said. 'It's amazing that she's still alive, this little one.'

She was a survivor. Alone and deserted, the cub had managed to stay alive in an environment full of enemies. She couldn't have been more than eight months old, standing only half a metre at the shoulder. It was impossible not to feel sympathy for the poor creature, but of course I knew that to interfere in any way was out of the question. There was no room for sentimentality out here. What had happened was nature's way, no matter how cruel it seemed to us.

Suddenly, Tristan's hand was gripping my arm. From the trees, swathed in dust and moonlight, came two gods. One of the males wore a crown of darkness, his thick mane undulating with every step; the other wore a crown of gold.

'King of beasts,' Tristan whispered respectfully.

My heart was beating like a thousand drums pounding an emergency message to a neighbouring tribe. Although I know it sounds ridiculous, I was genuinely worried it would attract the attention of the lions, who were now almost at the waterhole. I had the premonition of some sort of impending disaster, although I could never have imagined what would happen next.

The cub had sensed the presence of the two males before we had. She became agitated, sipping more quickly and becoming more vigilant. The golden-maned lion roared deeply and ferociously, all the might of the heavens in that terrifying sound.

With the benefit of experience, I can tell you now that I have only ever heard lions make that sound when they are making a kill. Most of the time lions don't actually roar; they *humph* to signify their presence to other lions. This is considerably less daunting than a full, harrowing roar, which to all self-preserving humans should not fail to evoke a kind of primitive, bowel-trembling fear.

The cub backed away, which was quite sensible I thought, and the males proceeded to lap at the shallow water, looking deceptively like overgrown house cats sipping from a bowl of milk.

The cub hung in the shadows, too afraid to move. Why she stayed there I failed to understand. Perhaps she didn't realise the danger she faced. Perhaps she did and was paralysed with fear. Maybe she sought the company of her own kind after being deserted by her mother and her pride. Perhaps she was still thirsty.

After the males had finished drinking, they loped over like arrogant schoolyard bullies to where the cub stood with her head lowered subserviently. The dark-maned lion walked up to a small mopane sapling. He stood over it, moving slowly above it so that the butterfly leaves stroked his underbelly, before wiping his scrotum over the bush to inundate it with his scent. Then he scraped his back legs into the soil in strong backward kicking movements. Dust flew up behind him. It was a blatant display of bullying that reeked of ostentatious machismo.

Then both males crouched on their haunches, less than a metre away, staring at her with cold eyes, appearing to destroy her spirit with their burning glare. She lowered her head even further and hunched her shoulders downward towards the ground. She seemed to be pleading for mercy, her small body a picture of utter subordination. She seemed to be *begging* to live. She knew her place in the pride order.

The males glared at her, wearing her down with their compassionless yellow eyes for what seemed an eon before they bore down on her. I inhaled sharply and held my breath. Time seemed to stand still. The two males roared simultaneously – the roar of the kill – and the three of them disappeared into a whirl of flying dust and chaotic, earth-shattering noise. In less than ten seconds it was all over. I breathed. The dust settled. The males settled back onto their haunches, their egos assuaged, watching the bloodied cub who had been in the middle of it all.

Their victim was somehow still standing despite being so viciously attacked. She continued to stand for another ten minutes, swaying slightly, at one stage dropping onto one front knee but then finding her strength to stand again. Eventually, after what seemed like an eternity, the males grew bored of her and sauntered off into the rapidly approaching night.

By now a deeper darkness seemed to shroud the waterhole. Perhaps a cloud had moved over the moon, or maybe a sinister shadow had fallen over my heart. The first excitement of seeing the lions had faded into a sombre, heart-wrenching mood.

'Bastards,' I spat, before replacing emotion with the scientific reason I'd been trained to adopt. 'I don't understand it! Why attack one of your own kind, let alone one of your future generation? It doesn't make sense.'

'That's Africa, man.' Tristan spoke gently, as if he too was deeply affected by this glimpse into nature's harsh reality. 'The web of life. You may as well get used to it. And the best thing about it is that the male with the blond mane was her father.'

'What?'

Deserted by her mother, now attacked so viciously by her father? It seemed contrary to all the laws of nature. The more offspring a male sired and helped raise to maturity, the more of his genes he was likely to pass on. It made no sense to kill your own progeny. They were your genes' life insurance.

'It's not uncommon in lion society,' Tristan explained. 'These two males are brothers. One will always be dominant. And it just so happens that blond boy here recently got his arse whipped by his black-maned brother, which means that all of the offspring previously sired by our blond friend are on borrowed time. Only the strongest genes can prevail and this cub had the genes of the guy that lost out. The mother knows it. That's why she left her cub. My bet is any other cubs of hers have already seen their arses.'

'What are you saying, that the female would desert her own cubs just because the dominance hierarchy of the males has changed?' I said, finding it hard to rationalise.

'I'm saying exactly that. She may seem like a bitch to us but it's the best possible decision she could make under the circumstances. She might lose these cubs, but if she doesn't conform

Guinea fowls peck for scarce pickings on a track dotted with piles of desert-dwelling elephant dung in Kaokoland, now known as the Kunene Region. Driving through this isolated, arid region in north-west Namibia was like taking a step back in time to a desolate land that was more like Mars than Earth.

Plains zebras congregate at the Okaukuejo waterhole in Etosha, along with springboks, kudus, oryx and a plethora of other creatures in the hot dry season, with temperatures frequently soaring above forty degrees.

Giraffes on Etosha's plains form a stylish silhouette against the pink hues of the setting sun.

In the late afternoon before rhino monitoring began, our team relaxed at Dolomietpunt in western Etosha, while an elephant bull drank at the waterhole behind us. None of us wanted to miss out on the opportunity to get up close and personal with one of the most endangered creatures on Earth. *From left to right*: me, Rainer, Stu and Werner.

A herd of black-faced impala close their eyes as a shower pelts down onto Etosha's parched earth. The wet season invigorates and refreshes not only the plants and the animals in the park, but also the people, who long for relief from the endless dry heat.

Naomi and I share a Savannah Lager in front of the tin box that I lived in for my first six months in Etosha. During October, it literally felt like we were being cooked in an oven, but Okaukuejo Research Camp soon began to feel like home.

Headed for Kaross, I stopped on the wide gravel road for a road block – a stoic black rhino. Annoyed by the intrusion, he gave my ute a solid charge, causing me to reverse swiftly back along the road to avoid invading his personal space.

Putting eleven radio-collars on female black-faced impala sounded like a piece of cake to me, but I couldn't have been more mistaken. Selecting, darting and finding the immobilised impala in the bush proved almost impossible. Brendan, Stu and I had an anxious six weeks ahead before all the females were finally collared, and I began the daunting task of tracking their movements with my research assistant, Savimbi. HANNAH HOARE

Naomi and I investigate the map on a section of the Etosha Pan in western Etosha, which is off-bounds to tourists. The possibility of getting lost or breaking down in western Etosha was always in the back of my mind. The likelihood of being found within a few days by park staff was fairly low, so I carried plenty of food and water just in case.

A little Himba boy stands outside his cow dung hut in the Kunene Region. Many Himba people still live traditional lifestyles, covering themselves in red ochre and subsisting in the desert with goats and cows.

An old Bushman woman in the Nyae Nyae Conservancy tells me through a translator how she is afraid of elephants. A woman was killed by elephant bulls in Nyae Nyae a few years before, so she had reason to be afraid. Cackling, she told me that if any elephants were to come along right then, I could stay, but she would run!

An elephant bull lifts his trunk to smell us at Tari Kora waterhole in Khaudum National Park, assessing whether we are friend or foe. Ironically, while people in Bushmanland are afraid of elephants, elephants also often live in fear of humans – a common problem in Africa where the ranges of humans and elephants are growing and now overlap. The safety of both species depends on finding solutions that allow people and elephants to live together in harmony.

to the pride order then it's only a matter of time before she's also vulture fodder, and there go all her chances of rearing any offspring in the future.'

My head was spinning with contrary thoughts. Torn between an instinctive emotional response and the scientific voice in my head that said this was a plausible reason for their behaviour, I was finding it difficult to focus. In the animal world, life was all about survival of the fittest. In this case, the maternal instincts of the lioness had been overruled by her own need to survive.

'It makes sense... I guess...' I began, but the words caught in my throat.

I knew that Africa was a harsh place but some things I could never get used to. Nature was extraordinary, but it wasn't always fair.

It was one thing to witness the ruthlessness of nature in the wild, but altogether another thing to put the animal you're supposed to be protecting in a position of risk. I'd obtained permission to dart and collar eleven female black-faced impalas in order to find out about movements and habitat use around the time of lambing. Immobilising any animal to put a radio collar on it has an associated risk, but I never expected the process to be so anxiety-filled.

I was sitting in a crudely fashioned hide, thirty metres from the Ongava Lodge waterhole. There was just enough room inside for three people to sit comfortably. In the hide, roughly hewn from the poles of mopane trees and camouflaged by their flapping, butterfly-shaped leaves, the temperature blasted past the thirty-five degree mark. It was September, the build-up to my second suicide season, at about eleven in the morning. The windless heat was a living, breathing force. Sweat soaked my faded green singlet.

Endless streams of mopane flies persecuted me, tiny black devils diving incessantly into every facial orifice, occasionally blinding me when one drowned in the mucus of my raw and bloodshot eyes. I was a million miles from nowhere, hounded relentlessly by heat and insects, and chomping at the bit for an impala to come in and drink so we could plant a fancy pink dart in its shoulder.

Any excitement I felt at the prospect was tempered by an anxiousness about how well it would go. Impala are neurotic animals by nature and I didn't want to lose one to stress caused by capture. This was an endangered subspecies and the last thing I wanted was blood on my hands. Then again, the information that I could glean from this study would be the first of its kind and would be crucial for the success of future translocations. We had to give it a shot.

Beside me Brendan reclined in his canvas deckchair. The Ministry of Environment and Tourism's young game-capture vet had immobilised common impala before in South Africa and wasn't looking forward to this.

'The bloody things *run*,' he'd told me. 'Impala... *humph*! If you don't get to them in five to ten minutes, the position of the head blocks the wind canal and they're dead.'

His professional concern only amplified my worries. Brendan had agreed to undertake the darting operation for free, both as a friend and because the ministry had a strong interest in supporting my research. Stu and Rainer and a number of other Ongava staff were keen to be involved too. After all, it wasn't every day that radio collars were put on black-faced impalas.

About a hundred metres away in the lodge, on the *kopje* behind us, a team of five volunteers, Ongava staff and my visiting family were watching through scopes and binoculars for any incoming impala. The middle of the day was usually when the impala came to the waterhole for a drink, but as the heat escalated, the risk of heat-stress related mortality increased.

We had a rough plan of attack, but as no one here had ever done this before, it was all a bit hazy. Brendan had explained that once he had taken his shot and the impala was darted, we all had to run like hell to get to the animal before it asphyxiated. Stu and Rainer had been told to run down from the lodge to help us transport the impala in a net that I would carry. We had to take the immobilised impala to a patch of shade near the hide that we called the collaring area. It was crucial that we hold the animal's head up so that it could breathe properly. At the collaring area, a canvas mat was spread out and a scale was hung over the branch of a mopane tree to take the impala's weight. Brendan said that he and I would depend on the instructions from the team up at the lodge with their better view from the *kopje* as we bashed through the thick, thorny bush in search of an immobilised impala. After he had made this vague plan clear to the team we had driven down to the hide to wait.

Brendan's foot rested lightly against the leg of my chair while he played with the fluorescent pink feather adorning the end of the dart. Neither of us spoke. A Dan-Inject dart gun leant against the wall of the hide, the hard steel of the barrel glinting intermittently in the dappled light. Brendan wiped a drip of sweat from the side of his eye, dislodging a mopane fly that was aggravating him in the process. We were both edgy and impatient. Waiting, waiting, waiting.

'Brendan, Brendan, come in,' Rainer's German voice blasted over the radio, startling us.

Brendan picked up the radio and answered, 'Ja, standing by, Rainer. What've you got?'

'There's a herd of about ten coming in to your right...several females in it.'

'Excellent. Make sure you keep an eye on the one with the dart, eh?'

'Got it. Over and out.'

Brendan opened his drug case and loaded up a dart. The drug he was using to immobilise my impala was the same one used to immobilise elephants and rhinos, only in much smaller doses. It is lethal to humans. It took one millilitre of M99 to knock an impala out. Apparently the same amount in a human could kill in less than a minute.

I barely moved an eyeball as Brendan leaned over to access the opening along one wall of the hide and perched his dart gun on the rough wood. He aimed at one of the females nearest to us, pointing the barrel at her rump. Seconds ticked by as an excruciating tension permeated the hide. Sweat beaded his forehead. He squeezed the trigger gently as the female impala he was focusing on leaned over the waterhole to drink, broadside to us so that her rump provided a fair target.

Crack! The dart flew through the air faster than I could see it and landed in the muscular hind leg of the female. The herd barked simultaneously in fright and all of them took off. I watched the female with the pink feathered dart in her rump leaping away, probably aware of the strange stinging sensation in her leg but unsure of its origin.

'Dart's in!' Rainer's voice boomed over the radio.

'Good,' Brendan replied. 'Keep an eye on her.'

From the ground height of the hide we lost sight of her within a minute. The thick bush beyond the clearing around the waterhole formed an impenetrable barrier, even with the use of our binoculars. We expected her to be down in about eight minutes.

After five long minutes Brendan radioed the team at the lodge, 'Can you still see her? Is she down yet?'

There was a tense pause while Rainer spoke to the rest of the team.

'We've still got her in our sight, ja... She's a bit wobbly, but still standing,' he relayed.

'Okay, let's go,' Brendan called.

Adrenaline surged through my veins. We charged out of the hide. The heavy net that we would carry the impala in was shoved awkwardly under my arm and it weighed me down. I couldn't keep up with Brendan, who was sprinting ahead. I jogged behind him in the midday heat, choking on the dust flung by herds fleeing ahead of us. The only way I knew roughly in which direction he had gone was by the bursts of radio communication in the distance.

'To your left, Brendan! She's down to your left and about fifty metres ahead of your current position,' I heard the radio echo.

Behind me, I heard Stu's feet pounding on the earth.

'Here, let me take that, Tams,' he said, puffing, and chivalrously took the net from me.

Without its dead weight I managed to keep up with the guys, tripping occasionally on the white calcrete rocks that formed much of the substrate.

I called out to Brendan, my voice husky from a dust-coated throat. In the thick bush, visibility was about thirty metres. A motionless impala was virtually invisible in this glaring, sharp environment. Sickle bushes with inch-long thorns that I'd seen penetrate right through the soles of shoes and into the soft flesh of feet blocked our path at every angle. *Wak-a-bitje* – wait a bit – bushes tore and scraped at my legs and arms, drawing blood from soft flesh and ripping in my clothes. But this was the least of my concerns. The female had been down for over ten minutes. I began to worry that we wouldn't find her and that she would die in the bush. The thought made me feel nauseous. Had this all been a big mistake? As if that wasn't bad enough, we all knew that there were lions in this patch of bush – they frequented the waterhole daily. The fact that both we and the immobilised impala were at risk from these predators nagged at the recesses of my mind.

A voice rang out from the bush about fifty metres from where I stood puffing like a steam train, scanning the bush. 'I've got her! Where's the net?'

I said a silent prayer of gratitude. Stu and I sprinted in the direction of the voice and found Brendan crouched behind the impala. She was comatose and he was holding up her head so that air passed freely to her lungs. I just about collapsed with relief, but there wasn't time to relax. Now the work began.

'Bloody males were trying to mount her!' Brendan said, grinning like a madman. 'She was wobbling around like a drunk and the rams were all over her. That's how I found her, because the rams were roaring like it was the rutting season!'

The impala's tongue lolled out of her mouth, a common response to the drug and a sign that she was totally out to it. She wouldn't wake up until Brendan gave her an injection of antidote, but the less time she was under the better. Rainer arrived just in time to help with the carrying and between the four of us we lifted the dead weight of the antelope onto the net. As the most physically weak of the team, I was allocated to carry her head and hold it up by the ears. That wasn't as easy as it sounds as we all jogged over rocks and logs, bashing through thorny acacias in the oppressive heat. The heavy breathing of the three strong men was synchronised with their thumping steps. I was surprised at the heavy weight of our female. I'd read that common impala only weighed about forty kilograms. This black-faced impala seemed heavier.

Finally, sweating profusely and feeling winded, we laid her gently on the canvas mat under the shaded tree at the collaring area. I crouched beside her, holding up her head, while Brendan took her vital signs.

'She's fine. Good shape,' Brendan announced. 'Okay, Stu, can you get going on the collar?'

While Stu undid the screws on the collar, I set to work with Mark, my uncoordinated, gentle giant of an English volunteer. I'd taken a bit of a risk taking him on for three months, never having met him and not knowing how he would adapt to life in Etosha. The dusty desert life isn't for everyone. But luckily, he was keen, competent and a lot of fun to have around.

Mark recorded notes on a data sheet as we took morphological measurements of the impala. The body measurements of black-faced impala had, to my knowledge, never been recorded, so this was important information for the national records. Brendan inserted a needle into a vein on the female's inner thigh and took some blood for genetic analysis. We knew that black-faced impala were physically different from common impala in other, less arid parts of Africa, but were they genetically unique? Taking blood would allow us to determine this.

Then, before the collar was put on, several of the blokes hoisted her up to be weighed on a makeshift butcher's scale. She weighed in at fifty-two kilograms, over ten kilograms more than the average weight of common impala in South Africa and pretty much the same weight as I was. No wonder we'd all been puffing as we'd carried her!

While Mark held her head up, Stu and I fastened the collar around her neck. The radio collar was the tan colour of impala fur, and a battery pack of the same colour, encased in fibreglass and plastic, hung from the bottom. If all went well, the batteries would last for two years, providing us with a good record of her movements.

'All done?' Brendan asked, wiping his brow, as Stu drove the final screws into the tough fabric with a screwdriver.

I racked my brain, trying to make sure there was nothing we'd missed. We'd taken measurements, blood and a nick of her ear for genetic analysis, removed ticks from her ears and legs and affixed the collar. All this had happened in a chaotic, stressful twenty minutes.

I nodded. My hands were shaking with tension as I placed them on the warm, sweaty fur of the impala's side. I felt her chest rise and fall with each rhythmical breath. Like all female black-faced impala at that time of year, she was heavily pregnant and due to lamb in a few months. I prayed that this experience hadn't affected her pregnancy in any way and that the collar wouldn't inhibit her activities.

We would be putting radio collars only on the females because I wanted to understand how giving birth influenced the females' behaviour. In an endangered species, building up the population depends on a successful recruitment of young, so the mothers' behaviour at the time around lambing was crucial to the survival of the entire population. How many females gave birth and how many lambs actually survived? These were questions that needed answering if we were to find out what limited the black-faced impala population.

Watching this particular female and feeling her soft fur, I felt profoundly close to her in a way that I never had when observing impala from a distance. I felt responsible for her. It was a feeling that would grow stronger the more days I spent walking with her and the other collared females, monitoring their movements, following the course of the herd and sharing the joy of the birth of their lambs. It was utterly unprofessional and unscientific for me to feel empathy for this animal, but the feeling wouldn't go away. Anthropomorphism is a heinous sin in scientific circles. The boys' club would have found these sentimental thoughts utterly hilarious, so of course I never said anything. But science doesn't have all the answers, and when it comes to understanding animals, even the experts are still barely scratching the surface. We are all limited in our understanding of animals by the very thing that makes us human – our human perspective.

Brendan leaned over and injected the antidote. A drop of sweat fell from his brow onto the sweaty fur of the impala. Now that our work was done, I breathed in the strong, musty smell of her and felt free to enjoy the moment. It was a privilege to be this close to a wild animal. I hadn't realised how beautiful these black-faced impala truly were.

'Everyone back!' Brendan said as he helped her up and pointed her in the direction of the bush.

About a minute later she was on her feet, albeit a little wobbly, as if drunk. She didn't seem aware of the collar around her neck. She leapt into the air in some sort of delayed fright response, then stopped, stood still and turned back to look at the gathering of people about ten metres behind her. The female's deep brown eyes surveyed us for a few seconds, as if she thought she was only dreaming, and then she bounded away and disappeared from our sight.

I must have looked a bit traumatised because Stu put his hand on my shoulder and said, 'No stress, Tams. It went well.'

He was right – it had gone well. I smiled at last. One down, ten more to go. The first collaring had gone so well that I figured the rest would be a piece of cake. How wrong I was.

I'd budgeted that it would take us two weeks to collar eleven females. Six weeks later, we were still collaring. Once again, I'd underestimated the intelligence of a prey animal. You'd have thought I'd have learned by then.

After the first female had been collared, it seemed as if the entire population of black-faced impala at Ongava had let each other know that there was danger at the lodge waterhole. When they came in to drink, they were skittish and twitchy; they reduced their drinking time and moved around so much that Brendan couldn't get a clear shot. Then, a week of violent winds increased their skittishness even more, and every time Brendan tried to get a shot, the dart was blown off target by the wind. Day after day we sat in the hide, waiting futilely for an opportunity to get a shot. Dart after dart flew into the vicinity of the impala and was either blown off course or the target animal moved out of the way. Sometimes the dart went in but the plunger must not have pushed any drug into the muscle, because we'd spend hours scouring the bush on foot and not find any sign of the female or the dart. Brendan was growing increasingly frustrated. My anxiety permeated my dreams at night, in which impala dodged darts like superheroes. My arms and legs were raked with cuts and scrapes from the thorny bush.

One time Brendan was resting his head on his gun case while I kept an eye out for impala; when I announced a herd's arrival, he got such a fright that he fell off his chair and kicked the gun up into the air. Somehow, while trying to regain his composure, he managed to get a shot in, but it was another case of the missing impala. She was nowhere to be found. We both knew they were getting the better of us. Brendan announced that he hated impala. This was a man who immobilised lions, rhinos and elephants for a living. His pride was irrevocably wounded.

But we couldn't give up, and to Brendan's credit, he stuck at it. One morning we drove over to Ongava from Etosha with my mother, father and younger brother, Davo, in tow. Davo, at sixteen, was having the time of his life and I knew he'd be back again after this trip. He'd caught the Africa bug, to my mother's despair.

As we rounded the bend to the lodge, the manager, Mark, pulled us up. 'I don't think you'll get in there today, guys,' he said, smiling. 'There are lions down there beside your hide.'

Mark joined Davo on the back of my ute and we moseyed around the rocky track to the waterhole, keeping a sharp eye on the bush for any movement. We spotted a lioness first, her face painted with the fresh blood of the dead oryx she was feeding on. They'd killed it just thirty metres from the hide and had settled in to feast for the morning. Another female and a male lion were nearby, resting in the shade. The lioness at the kill glared at us as we watched her. It looked like the others had fed and she was only now getting her fill.

On another day, after a dart had bounced off a female's rump yet again, Brendan was hurriedly reloading a dart with M99 when it flicked out of his hand, bounced off the gun case and a splattering of the lethal drug splashed above his eye.

'Fuck!' he exclaimed. There was a tremor of fear in his urgent command: 'Tammie! Give me that water!'

I didn't realise what was going on and passed him my bottle of water, thinking how rude and demanding he was being. He was suddenly sweating profusely.

'Oh my God, did you get some of that on you?' I felt the blood drain from my face.

He was dousing his face with water, trying to wash the drug off and hoping that none of it had entered through the permeable membrane of his eye. It took less than a millilitre of M99 to have fatal consequences for humans.

'Ja, above my eye…' he said.

I noticed that his hands were shaking.

'Shouldn't you give yourself the antidote?' I said, wondering why he hadn't already done it.

'Maybe...' he replied, and then to my astonishment, began loading another dart. 'Is the herd still there?'

'Brendan! It's not worth the risk. Give yourself the antidote, man!' I cried.

A minute had passed since the M99 had splattered on his face. Sweat bubbles lathered his forehead. I felt sick. What would I do if he passed out?

'I feel okay,' he said, with characteristic bravado. 'Don't worry...'

He leaned out of the hide and, amazingly, took a perfect shot at an impala. The dart went in and stayed in. Five minutes later, Stu radioed from the lodge that she was down. I was ready to run out of the hide but I didn't want to leave Brendan, just in case he showed any symptoms.

'Aah, what the hell...' he said, then quickly jabbed himself in the arm with a needle filled with antidote. I thanked God that the African male's bravado could be dropped for a moment of sanity that may have saved his life.

After that day, we realised that darting from the waterhole wasn't working effectively enough, so we decided to change tactics. From the back of the vehicle, with Rainer driving, Brendan and I stood and attempted to use a net gun. This commando device looked like something out of *Terminator*. When we got within twenty metres of the target animal, the idea was to fire a pre-packed, four-by-four metre orange net over the animal to catch it. Unfortunately, most impala had us sussed by now and wouldn't let us get close enough in open country to get a shot at them. When we finally did get a shot, it was a dud bullet and misfired.

With the Dan-Inject gun that we'd been using in the hide, we could afford to be a little further away in the back of the ute, but estimating the distance between us and the animal was the hard bit. The gun had to be set to the exact distance in order to be accurate.

'Should I take the shot?' Brendan said, lining up a female in his sights as he leaned on the back of the moving Hilux.

'Yep. Go for it,' I said and gently tapped on the roof.

Rainer slammed on brakes and dust rose up around us in a choking cloud. The impala disappeared into the white talcum-powder shower momentarily and then reappeared. Brendan took the shot, leaning on the roof of the ute. I watched the dart fly into her neck and then bounce off. This had happened several times, when the exact distance wasn't estimated correctly and the power wasn't set perfectly. Brendan shook his head and swore. I could tell he was growing to hate impala more and more by the day. We still had only one impala collared after three weeks.

I was becoming frantic. I'd fundraised long and hard for these radio collars and the thought of not being able to get them on the animals was genuine reason for anguish. I'd even been featured in Brisbane's *Courier Mail*, with glamorous photographs taken of me in full khaki attire, trying to raise funds for the collars. The media attention had resulted in some good sponsorship and led me to meet Warren Tapp, a Brisbane businessman with a kind heart and a generous spirit, who would continue to support my work in Africa for years to come. I felt indebted to my sponsors and the possibility of letting them, my supervisors and myself down didn't bear thinking about. We had to get the collars on somehow. We decided to give the impala and ourselves a break for a week, while we all caught up with other work.

When we got back to it, Brendan upped his dosage of M99 to 1.2 millilitres and Nad lent him a different dart gun. It was newer, smaller and had more power... and, it worked!

'She's over!' I exclaimed.

The female had wobbled a little, and then toppled onto her side on the powdery dust. I launched myself out of the back of the Hilux before it had stopped and began running towards her.

'What? Over what?' Brendan exclaimed, finding it hard to believe that we'd actually managed to get something right. 'Over the road?'

'No! She's over! She's down!' I exclaimed on the run.

I was all grins as the female, equipped with a new collar, bounded away into the bush after an efficient, twenty minute collaring operation.

With the new gun and increased drug dosage, things flowed better than they had before. Even so, by the time we'd put our final collar on, we were all ready for a holiday. The endless anxiety of not being able to immobilise impalas, then the risk of not finding them if they were drugged, had drained us all. But for me, the hard work was only just beginning. I was about to start radio tracking.

WALKING WITH THE IMPALAS

Tristan stood on the back of my car and held the radio-tracking antenna high above his head. His metallic sunglasses reflected the white sun, shooting a blinding ray off their mirrored lenses. The receiver, a small box protected by a canvas and plastic covering, hung over his broad shoulder by a canvas strap. He turned it on and the squelch roared into life. I couldn't hear any beeps until he turned the antenna to face west and then...*beep...beep... beep...* Ever so faintly, the beeps indicated that the impala we were tracking was west of us. We got back into the car. I drove, while Tristan and Savimbi, my new research assistant and tracker, a happy-go-lucky Himba man in his early twenties, stood on the back. I'd decided to take Savimbi on because of the threat of lions and rhinos at Ongava and the fact that we wouldn't be armed when walking in the bush every day. I needed a second set of eyes.

Every ten minutes or so we got out of the car and checked the direction and volume of the beeps. Soon the beeps were so loud it sounded as though we were right on top of her.

I stopped the car.

'Let's have a look around,' I suggested.

The three of us split up in different directions, but still within sight of each other. We'd been driving around the same small area for over an hour; if the female we were looking for was still alive, she wasn't moving very fast. I knew it didn't bode well.

'Look for a collar or a pile of bones,' I yelled.

As we searched the rocky terrain, I was slightly distracted by Tristan's presence and finding it a little difficult to concentrate. He'd volunteered to come and help me track impala on his day off. I was hyperconscious of the feeling that he wasn't just interested in the impala. I was aware of him watching me in a very different way to the boys' club in Etosha, and was surprised to find myself a bit flustered around him.

I had to admit to myself that I was impressed with Tristan's understanding of the bush. Clearly he had developed a deep respect for wildlife when growing up at Waterberg, but he was also trying to get into the minds of the animals in order to understand why they did what they did. It was something Roger had always told me to do. 'Watch their behaviour,' he'd said. 'Think about why they do what they do. You can read the bush like a book if you just know how to.' He proved it too – he would arrive at a waterhole and within a minute he'd know what species, how many and what sex and age had been there, whether anything was injured or if fighting or a mating had occurred there.

Tristan was, in many ways, a younger version of Roger. He had the same fiery temper and soft heart, the same love of the bush and ability to read its signs. Like Roger, he had a rebellious nature and hadn't finished high school. As a result, he'd developed his practical skills. He'd built his first car at the age of fifteen. That was when I'd been learning to drive.

There was something infinitely attractive about this man, with his broad shoulders and muscled biceps, something raw and untamed about him.

A call from Savimbi rang out of the bush, about thirty metres away, awakening me from my reverie.

'Eh-he!' he called.

He was standing beside a pile of bones. All that was left of our collared female was two hind legs, devoid of meat, and the white bones had been scattered across the area by the undertakers of the bush – the vultures, hyaenas and jackals. Tristan bent over and picked up the collar. Amazingly, although it was covered in a sticky conglomeration of blood, mud, hair and dried predator saliva, it hadn't been chewed up too badly and the transmitter was still working. Thankfully, we would be able to put it on another impala.

These were the remains of a female we'd christened 'Grandma' during the collaring operation. She had weighed less than the other females and her gaunt face and bony body had borne the strain of old age. Under her tail had been a strange mass of white risen marks that Brendan had never seen before, a sign of old age, he'd guessed. It was impossible to tell how she had died. No one could say for sure if having a collar had made her more vulnerable to predators or whether she had already been on the way out anyway because of her age.

At the time, having just started the study, I was very worried about this. Radio collars generally don't influence animals' activities, but it is something that researchers always have to be aware of. To my relief, two years later all eleven females (including the one we later collared with Grandma's collar, who was in such good condition that we christened her 'Miss Namibia') were still alive and each of them gave birth to a lamb, so I could safely say that the collars caused minimal disturbance. The idea of this study was, after all, to help conserve the subspecies, not eliminate it.

During this work I found it reassuring to have the expertise of Savimbi, my master tracker.

Savimbi was a couple of years younger than me and had grown up in the area formerly known as Kaokoland. He hailed from the beautiful ochre-painted, copper bracelet donning tribe known to some as the 'lost tribe of Africa', the Himbas. His people lived in mud and grass huts and herded their goats and cattle on the banks of the mighty Kunene River, which formed the border between Namibia and Angola in the north and further south into desert lands where no one else could survive.

Savimbi spoke a little English, but was more comfortable in Afrikaans, so we alternated between the two languages, helping teach each other. By now I could speak and understand enough Afrikaans to hold a disjointed conversation. Savimbi could also speak Otjiherero, because the Herero and Himba languages were very similar, both ethnic groups having evolved from the same root race. The word Ovahimba means 'beggars'. The tribe that came to be known as the Himbas came from a group of Herero cattle herders who were forced to flee their fertile homeland by Nama warriors in the nineteenth century and so came to live in the arid expanses of Namibia's north-west. In modern-day Namibia I'd noticed that the Herero had adopted a relatively modern way of life, many as commercial cattle ranchers, while the Himbas continued to eke out a traditional existence in the desert.

However, Savimbi considered himself a Himba of the modern variety. He wore western clothes and scorned the idea of dressing the traditional way. With nothing more than a primary school education, he'd taught himself the basics of five languages – German, English, Afrikaans, Damara and Oshiwambo – in addition to Herero and Himba. He'd come to Ongava on a short-term contract to help with maintenance work, which is how he met Tristan. When the two men had been fixing a pump near the waterhole, lions in the near vicinity hadn't bothered Savimbi in the slightest.

'I am not afraid of lions,' he had said, laughing, when Tristan had enquired about this.

I liked the idea of working with someone who wasn't afraid of lions because his main job would be to watch my back. He also had a laugh like a hyaena, which never failed to make me smile.

On our first day of tracking impala together, I decided to drive up to the southern *kopjes* of Ongava to see if any of the collared females were using that area.

'Is this the right road?' I called out to Savimbi, who was standing in the back of the ute to get a better view.

'Ja!' he replied.

I kept driving, feeling that we were heading deeper and deeper into nowhere. The track tapered out into not much more than a game trail.

'Savimbi? Are you sure this is the right road?' I called, sticking my head out the window, careful to avoid having my neck ripped off by an increasing number of overhanging thorny branches as the track narrowed.

'Ja, okay! Ja, okay!' he called back happily.

'Did you see the turn-off?'

'Ja, okay!'

'No, Savimbi, I'm asking you a question,' I said, growing a little frustrated.

'Okay, ja!'

Fortunately, Savimbi's English comprehension improved quickly.

I had to admit that having Savimbi around gave me a huge sense of security. He had the eyes of a hawk. Often his traditional tracking skills found the collared impala before my sophisticated, modern radio-tracking gear did. Our goal was to find all eleven females every day over a period of five months, but at the start we would be lucky to find half of them each day. I walked behind Savimbi at a soldier's pace until we grew close to the herd, at which point we slowed down to a stalk. Checking the wind constantly, surveying our environment with our ears and eyes, we trundled through the thick bush until we were close enough to see the female we were looking for. It took me a while to get the hang of stalking in the bush again, whereas Savimbi was an old hand. But soon we were working as a cohesive team, spotting most of our females each day. *Beep... beep... beep...* The constant sound of the radio-tracking equipment punctuated my days and haunted my nights: even my dreams beeped incessantly.

Savimbi always found humour in the behaviour of animals, which shed a different light on the way I looked at them as a budding scientist. I'd often hear him chuckling to himself for reasons that weren't at first apparent to me. Animals *amused* him. When the impalas outsmarted us he would chuckle quietly, shaking his head and tut-tutting, as if he and the impalas were on a daily mission to beat each other. I have never worked with a tracker who enjoyed himself so much in the bush.

He got a huge kick out of being able to spot the collared impala before I did with my tracking devices. As I followed behind in his footsteps, holding the antenna high above my head with the receiver beeping away, he'd suddenly stop, smile and point. Then he'd look at me, grinning with amusement at the fact that he could see the impala in the bush and I couldn't even see it through my binoculars! Long after I'd lost my sense of humour from the heat and dust and skittish impala, he'd still be grinning.

One day, as we squatted on the ground, watching a herd and looking for a particular collared female, we heard a burping sound. We both looked up and saw an old blue wildebeest bull walking towards us, oblivious of us sitting in the dirt in its path. Savimbi's face ripened into a wide grin. The wildebeest burped again, his dopy head nodding as he traipsed along on his path like a senile old man. When he was about five metres away from us, the black whiskers on his shaggy chin twittering like twigs in a bird's nest on a windy day, he suddenly realised we were there. Hilariously, he skidded to a halt, stood stock-still as if stunned for an instant, before bolting in the opposite direction. Savimbi and I burst out laughing. Savimbi pointed a finger at his own head and raised his eyebrows, gesturing that this wasn't the smartest animal he'd ever encountered.

Another day, while walking through the bush, we heard the rhythmical grunting of two animals simultaneously. *Grunt, humph, grunt…* It sounded like two pigs on a military march. In fact, it was two militants, two honey badgers marching through the bush, their white and black fur ruffling as they grunted, intent on their mission.

If I could aspire to be like any animal in the African bush it would be the honey badger. I recalled another time in Etosha when I'd heard a scratching sound coming from somewhere ahead on the path. Stepping quietly, I saw that it was a honey badger, digging with his strong claws into the base of a termite mound, looking for food. I had squatted down and watched in silence.

The honey badger hadn't seen me and the wind was in my favour. The tough little animal continued to dig with his robust arms, grunting as he flung dirt out of the hole. Behind him, a pale chanting goshawk strutted, his riotously orange beak and legs in stark contrast to the subtle, earthy colours of the dry bush. Suddenly a small rodent leapt out of the hole where the honey badger had been digging. Grunting furiously, the honey badger tore after it, but with lightning speed, the goshawk beat him to it and carried it off into a nearby shepherds' tree. The honey badger didn't give up. Instead the tenacious animal moved on to another hole and began digging until another mouse escaped. This time the prize was his. His persistence had paid off.

Honey badgers, determined to the very end, live their lives as though there is no tomorrow, and they're afraid of nothing. I've heard of a honey badger being flung around by an elephant, still sinking its teeth into its trunk, even after being trodden on a couple of times. This is an animal that eats cobras for a living. Many people visiting Africa consider the lion to be the king of the jungle and the elephant to be the wise old elder. If this is the case, then the honey badger is most certainly the greatest warrior of the African bush.

Some days of radio tracking brought with them unexpected rare sightings; other days we simply got the job done, tracking the impala until it was too hot to stay outside. It was sweaty, heavy work and, at times, downright dull. But Savimbi always found novel ways to entertain me.

On one occasion he noticed with rapture a glistening, golden sap on the peeling bark of a white-thorn acacia tree. He yanked the hard sap off and began chewing on it. A warm, happy glow spread across his face as he sucked on his prize. He passed me a piece and encouraged me to eat it, saying, 'Bush lolly.' Tentatively, I bit off a small piece. It tasted like butter menthol mixed with eucalyptus, sweet as a lolly, just as Savimbi had said.

In the middle of the day, when it was too hot to track, I snacked on tinned tuna, beans and bread, typed data into my laptop and then had a midday siesta. By three in the afternoon, Savimbi would always be waiting for me, ready to go again. In his time off he crafted Himba bracelets made from white poly-pipe, hand-carved with a knife and painted bright colours. The intricately crafted bracelets danced with African designs and sold for the equivalent of ten Aussie dollars at the lodge. This supplemented what was then considered the good local daily rate of five Aussie dollars that I was paying him. With his gentle way and quiet patience, Savimbi was not only a fine tracker but also an excellent artist and a businessman. I couldn't believe my luck in having not one, but two fabulous men in my life.

When I'd first moved to Ongava for this study, I'd moved into a room in Rainer's house. Then, on one morning, I woke up under my white mosquito net to find flowers on the table beside my bed. Each morning, before I woke, someone would pick native flowers from the bush on the way to work, sneak into my room and leave the flowers there for me to wake up to. They were the first things I saw in the mornings. I am a light sleeper, so I was surprised that the person managed to sneak in without waking me.

I had a feeling I knew who it was, but I wanted to catch him in the act. And on the third day, I was awake when the man in khaki tiptoed in.

'Tristan!' I exclaimed.

He burst out laughing and so did I, both of us feeling a bit embarrassed.

'Thank you,' I said, feeling genuinely flattered, but a little shy.

He pulled the net aside and gently kissed me. Then he stood up, his intense gaze still fixed on mine, and walked away, smiling with his eyes, leaving me in a state of pleasant shock and wondering if I was still dreaming. He'd won me there and then. The sexual chemistry between us was off the radar, beyond anything I'd ever experienced before. I no longer cared if we had a future together, given that I was leaving Namibia in a year. I just wanted to explore the possibilities and see where it led.

A month later, Tristan convinced me to move in with him at the opposite end of the farm, an old farmhouse called Leeuport. I was only planning to be there for a few months while I was radio tracking at Ongava, so I didn't take it too seriously.

Often, after a day's work, we would walk, hand in hand and without speaking, to a nearby waterhole and watch the sunset. In Tristan I had met someone who was perhaps even more inspired by nature than I was, and it had nothing to do with mastering nature as it did with some men in Africa, and everything to do with a kind of deep reverence and respect. His passion for wilderness was contagious. Since he'd moved to Ongava I'd seen him thrive on being back in the bush. Physically, he became leaner, more muscled and developed a deep tan. It seemed both of us were growing into our skins by being immersed in nature, and as that happened we realised that a strong bond was forming between us.

I knew I had to leave Namibia in a year to return to Brisbane and write up my doctorate, and I didn't want a long-distance relationship, but Tristan was very persuasive. For the first time since I'd moved to Etosha, I actually felt like a woman, worthy of a man's desire, rather than a desert cactus.

Leeuport, meaning gateway for the lions, was the original tin-roofed farmhouse at Ongava, from the days when the owners had run cattle. Cobwebs draped like curtains off broken windows. Owls had built a nest in a hole in the roof and the walls had browned from years of wear and rain. Attached to the old house was a flat and this is where Tristan had settled himself. The place hadn't had anyone in it for many years until he moved in, cleaned it up and repainted everything the colour of sunshine.

To be honest, the flat was a bit of a dump, but we overlooked that. Besides, researchers don't need luxury. I just needed a cool place to chill out when the heat got too much. Large exotic trees provided much-needed shade in the middle of the day. We turned an old hay storage container into a plunge pool, painting it in bright African designs. Once a day Tristan ran the generator for an hour to pump water to the tank, filling the pool and providing water for showering. We had hot water, heated with solar panels, but some days the water coming out of the pipes from the tank was hot enough to brew a tea bag in.

The flat was surrounded by a low cattle fence which Tristan hooked up with electricity to keep the lions out. In the moonlight, after a sundowner and a dip in the pool, we listened to the lions roar from the waterhole a few hundred metres away and watched white rhino, red hartebeest and oryx feeding on the periphery of the fence.

One thing I hadn't realised about wildlife research was how *boring* it can be. When it's forty degrees and the sun is boring a hole in the back of your neck, your legs have been ripped to shreds from wait-a-bit bushes and your calves are aching from the strain of walking up and down *kopjes* without a break, and then the impala you've been tracking for the last two hours takes off before you can record her position, it's enough to give you the screaming horrors. Sometimes, in the lead-up to the rains, I just wanted to lie down under a big, shady tree and dream about flowing water. Instead, I settled for the thought that Tristan would have a cool bath filled with bubbles and surrounded by candles waiting for me when I knocked off tracking for the day.

After two months, the radio tracking was so monotonous that I was becoming blasé. That is, until the week the predators gave us a number of serious wake-up calls.

Savimbi and I were walking up a *kopje* near the old airstrip, with the receiver beeping away as it always did, the antenna held above my head.

'This way,' I told Savimbi, based on the direction from which the beeps were loudest.

We continued uphill, climbing over rocks and steadying ourselves on trees as we went. It was just after seven in the morning, but the day was already heating up above thirty degrees. And this was the cool part of day. It would take us about twenty minutes to get to the top of the *kopje* where we knew the impala liked to browse on a long, wide plateau. I felt my calves and quadriceps tighten, readying themselves for another steep climb. We had been doing this perhaps four times a day.

Suddenly a radio-collared female appeared to fly off the *kopje* above us like a lightning bolt. She raced downhill, a flash of rufous fur, narrowly missing us, her eyes flared. She hadn't seemed to see us. She was running as if her life depended on it.

'Hyaena!' Savimbi exclaimed.

I followed his gaze to where two spotted hyaenas were running down the hill in their swift but awkward gait, chasing the impala and coming straight for us. Savimbi gestured for me to stay still. He was smiling, enjoying the entertainment as always. There wasn't enough time for us to move anyway, because the hyaenas were gone in a matter of seconds in pursuit of their prey. They also hadn't seemed to notice us as they loped downhill about fifty metres away from us.

'Aahh,' Savimbi commented knowingly. 'I think there are lions here today.'

'Why? Do you see spoor?' I asked, a little alarmed.

'No,' he shook his head and then smiled, 'but they are here…'

I knew better than to question Savimbi's instincts, but we had a job to do and we had to get on with it. Knowing that lions were in the vicinity, we began walking up the *kopje* with renewed vigour and vigilance. I was unarmed, as usual. Fastened onto my shorts was what we called my penny banger, a device that looked like a biro that let off a very loud bang resembling a gunshot. To date, I'd never had to use it. In fact, I didn't even know if it worked.

An hour later we had found our herd and, in it, two of the collared females. We were now recording up to nine collared impala a day and were pleased with our efforts. We hadn't encountered any lions, so we padded off in search of our next target without having had a chance to test out Savimbi's assertion that he wasn't afraid of them.

The next morning, in the same place, Savimbi heard an animal being killed by lions as we trekked up the *kopje*. I didn't hear a thing.

'It is crying out. I hear it,' he said, astounded that I couldn't hear it too.

'Where is it?'

'Over there, on the other side of the airstrip, maybe one kilometre away... maybe not so far.'

About five minutes later, as we recorded the vegetation where the impalas had been foraging, the piercing recoil of what sounded like a gunshot shattered the whirring of the cicadas. It was close to us. Savimbi was about fifty metres away from me, looking for the herd. In the thick bush, I couldn't see him. I was about to call out to him, when an English woman's voice rang out on my handheld radio.

'NJ! Are you all right?' the woman asked, sounding alarmed.

There was a pause, during which the hackles on the back of my neck stood up on end. My heart almost stopped. A gunshot is the last resort when a guide is being seriously threatened in the bush. NJ, one of the guides at the lodge, was notorious for getting up close and personal with lions. This was why his guests loved him. He was the original adrenaline junkie and, with his rugged looks, long sun-bleached hair and deep tan, he epitomised the gung-ho African safari guide. This time, had he gone too far? I felt my face blanch with fear. Where was he? The woman on the radio must have been one of his tourists.

At last NJ's deep voice came across on the radio, sounding shaky. 'I'm fine. I'm coming now.'

Immediately I got on the radio and called him. 'NJ, what's going on there? Was that your shot?'

He answered, 'Ja, Tammie. What's your location?'

'I'm up on the *kopje* beside Margo, about halfway along the airstrip. Where are you? Are you all right?'

Then the radio began to break up. I caught only a word of NJ's response in between the static: lion.

Later that night, sitting around the bar, NJ told the story of what had happened. The golden light and shadows created by candles in glass lamps flickered across his face as he shook his head.

'I didn't know there was a fresh kill there. I'd never have gone in if I'd known that. I thought it was an old kill...' NJ was still visibly shaken by the day's episode.

I recalled how Savimbi had said that morning that he'd heard a kill being made nearby.

'I found the kill... Then next thing, the lioness came out of nowhere! She was moving so fast I didn't even have time to move. Man! She *flew* past me, coming in close, almost touching me. I didn't even see her coming. She must be the female with cubs. Then, next thing, she came back at me again, just as fast. I let rip with the flare, just to get out of there. That made her back off... Man, was she pissed off! And that's when I got out of there.'

So it hadn't been a rifle shot. NJ had let off his penny banger. He'd been very close to being mauled and he knew it. He also realised that he'd put us in danger too, because Savimbi and I had been on foot a few hundred metres from where he'd been charged. Walking into a pissed-off lioness was not something that I looked forward to on the best of days.

Savimbi, on the other hand, did look forward to the day when we would walk into lions and found my obvious trepidation quite amusing. When you are walking every day in country where there are lions and you are walking with the animals that the lions eat, it's inevitable that you will come into contact with each other. I knew the day would come. When it did, it was the impalas that gave us the warning.

Savimbi went ahead, as always. Sweat dripped from my forehead and stung my eyes. We'd found three collared females already and it was only midmorning. We were making good progress. Now, as we scaled up and down another *kopje*, I wasn't thinking of impalas or anything else that I should have been thinking about. I was in another world, dazed by the sun, blasé from the monotony of tracking, tired from the daily trekking. My brain had switched off. It was a very stupid thing to do. Luckily, Savimbi was on the ball.

The beeps of the receiver indicated that we were close to the female we were looking for. I knew from experience that we would find her within the next hundred metres. I jostled with the volume, fine-tuning our direction. Suddenly we heard them.

The staccato barks of the herd shattered my walking lull. A dozen impalas barking simultaneously is not a terrifying sound, but it's enough to make you wake up and listen. Usually when I heard that sound I knew the herd we were tracking had detected us. Today there was no way they could have sensed us. The wind was in our favour and we were camouflaged behind a bush. There was a valley in between them and us.

It wasn't us the impala were alarmed about. It was lions.

The fully-fledged roars that emanated from in front of us seemed to be drawn from my worst nightmare. If you can imagine what a bunch of fighting alley cats sound like, deepen the sound and increase the volume, you have the sound of lions making a kill. It chilled me to the bone. Goosebumps rose on my skin in sharp contrast to the sweat on my brow. I felt genuine fear. The lions had killed an oryx between us and the impala, which made them less than a hundred metres from where we stood, frozen in fright. Well, that is to say, *I* was frozen in fright. Savimbi was grinning.

Finally I found the courage to swear. Savimbi didn't say anything. He looked at me and then began walking in the direction of the lions.

'Savimbi!' I cried. 'Come on, man! We're not walking in there now. We don't need to see the impala that much!'

Savimbi's face showed his disappointment. 'But I want to see them.'

I could tell by the look on his face that this gentle Himba man was indeed unafraid of lions. They fascinated him, as did all animals. I could see he was contemplating the risk and I began to wonder if he wasn't going to go up there alone. Then he laughed to himself, shaking his head, as if thinking better of it, and began walking with me to the *bakkie*. I was relieved to make it back to the ute, but Savimbi kept glancing back up the mountain, hoping for a glimpse of the great cats.

We had a lot to thank the impalas for – the immaculate timing of their alarm barks had stopped us walking into a lion kill.

Savimbi summed it up perfectly, chuckling to himself as we got into the vehicle. 'Ach, the impala, he is knowing this thing… He is telling us about the lion, that one.'

FORCES OF NATURE

'Get back! Don't come any closer! There's a snake under the bed!' Tristan yelled, his face contorted in alarm.

After a morning of radio tracking, I had found Tristan standing on the bed, one hand gripping his forehead, which glistened with sweat, the other flailing around in the air. Unnerved by his outburst, I tentatively bent down to look under the bed, expecting to see a massive snake. But there was nothing there. The bedroom was dark, even though it was close to eleven in the morning. I grabbed a torch and peered under the bed.

'I don't see any snakes... Are you sure?' I asked.

I wondered whether the snake had slithered under a cupboard and out of sight. But something wasn't right here. It was uncharacteristic of Tristan to be perturbed by snakes. This was a bloke whose school friends had called Snake Man because of his passion for catching both harmless and venomous snakes. His thumb bore the scars of one unfortunate encounter with a newly discovered species of adder. The tip of his thumb had had to be cut off and a long scar testified to the effects of the venom. But he still loved snakes.

'Just don't come any closer!' he yelled.

He didn't need to yell – I was only a couple of metres away from him.

Suddenly he sank onto the bed as if exhausted, his brow furrowed in pain.

'My head... my head...' he whispered. 'Don't... It's here somewhere... I heard it.'

I rushed over to the side of the bed, forgetting all about the snake. I'd never seen him in such a sorry state. Although I'd never seen anyone with advanced malaria before, I knew that hallucinations were a common symptom of the dreaded disease. When I'd left him in bed that morning, he'd said he was going to sleep it off. He'd had typical flu symptoms and a headache at that stage. But four hours later, he was hallucinating about snakes under the bed. He was in such agony that he could barely move. I had to get him to a hospital and fast. The nearest one was in Outjo, an hour-long drive.

'Right, I'm taking you to the doctor,' I said.

In a heart-wrenching voice that bespoke the pain he was in, he replied, 'I don't think... I'm going to... make it.'

'Well what are you going to do then, Bushman? Just stay here and die?' I asked, trying not to shout.

I parked my *bakkie* next to the door of the house and, with his arm over my shoulder, helped him to the car. I drove as slowly as I could to avoid jolting over the rocky track, but Tristan still seemed to feel every bump as though it were a vicious blow to the head. When we got to the highway, I pelted along at 140 km/h. The national speed limit in Namibia is 120 km/h, but this was an emergency. I prayed that a kudu or steenbok wouldn't jump out in front of the car. Tristan would later say it felt like the longest drive of his life. He slipped in and out of consciousness all the way.

In Outjo a grey-haired Afrikaans doctor attended to him in the state hospital. The single-storey hospital was spartan but clean. I must admit I was pleasantly surprised, having heard atrocious stories about rural African hospitals. Outjo, being such a small town, didn't have the congestion of patients that other hospitals in Namibia did. In fact, I think Tristan was one of only about three patients in the place.

He was barely aware of what was going on as the nurses jabbed him with needles and put him on a quinine drip. I held his hand, trying to be of some comfort, but as the long metal needle pierced his skin and sunk into muscle I had to turn away. I felt the nausea that precedes a blackout and a star-filled darkness began to cloud my sight. It was suddenly very hot in the small room. Quickly, knowing the signs, I sat down and put my head between my knees. I waited for the nausea to abate and my vision to clear.

I have never been very good with needles. Ever since Grade Nine, when I, along with my fellow science students, pricked my finger to test for my blood type as part of an experiment and I passed out on the lab floor, I've known I could never be a medical doctor. I would face a charging lion or an elephant bull in musth any day rather than have my arm jabbed with a needle. This is a bit of a problem when you're addicted to Africa and have to have regular injections for typhoid, yellow fever and various types of hepatitis. I've lost count of the number of times I've had to tell the nurse that I'm about to black out. It's not only when I'm getting needles myself; I can't even deal with animals getting them.

So, while I was trying not to black out, Tristan was burning up and he was shaking with the excruciating pain in his muscles. When you have malaria, I'm told, you really feel like you are going to die. Plenty of people do, of course. The mosquito that transmits malaria kills more people in Africa than any other animal, including hippos.

That night, and for the next two, I tried to sleep in the squeaky, lumpy single bed beside Tristan, who was strung up to a drip. Sharp wire coils in the sunken mattress stuck into my back as I tossed and turned, checking on him periodically and trying not to worry. Although he was in hospital, he wasn't out of danger yet. He got up and stumbled to the toilet at least twenty times each night because the quinine caused severe diarrhoea and nausea.

Three days later I took him back to Ongava, but it was over a month before he had his strength back. I was just pleased that he was still alive.

It was January and the rains had come. Sweet, pure, unadulterated rain. God's most precious gift, falling from the heavens, brought healing and replenishment. Sitting on the cement steps at Leeuport, I closed my eyes and breathed in deeply. The smell and the sound of the rain on the tin roof re-energised my soul. The magical, earthy smell of rain on dust filled my heart with joy and vitality. There is no greater smell or sound in the world than this, I thought. For a few minutes the rain pelted down with such force that it sounded like applause, a thousand clapping hands, nature's own standing ovation.

With the rain would come new life, the next stage of nature's cycle. Fresh, green grass would spring from the blackened earth where widespread bushfires had marked their territory, filling the land with food for the hungry bellies of Etosha's grazing animals. I knew that life was about to take a new turn for me too. It was now only a few months before I had to leave Namibia and return to Australia. It was a matter of necessity, but it was going to hurt.

I *was* happy. I was living my dream. I was a part of the Etosha community and had gained some wonderful friends. I had a sexy boyfriend with an accent who loved the African bush as much as I did. I never wanted this time to end. The once brittle grasses of Ongava and Etosha had transformed, sprouting glorious new leaves. The plains were so green that it almost hurt your eyes to look at them. Water abounded in the vegetation and formed short-lived puddles on the roads. Baby springbok and wildebeest frolicked playfully on the plains, their stick-like legs providing unsure footing for soft hooves on muddy turf. Giant slimy bullfrogs emerged from months of hibernation in dark earthen caverns, croaking obnoxiously, taking advantage of the puddles to engage in a short-lived mating frenzy. There was excitement in the air and it spread contagiously throughout the natural kingdom.

To my delight, after a gestation period of about five or six months, the black-faced impala females gave birth to lambs. With the onset of the rains, they all dropped their babies within a couple of weeks of each other. This is a strategy to maximise the survival of vulnerable young, in that by birthing all the lambs at once, the predators become satiated, so can't eat them all. The mothers move away from the herd to give birth and leave their lambs 'lying out' alone or in a crèche of lambs while they go away to feed. It's the equivalent of day care. A herd is conspicuous and attracts more attention than a single female or lamb.

Each female stayed away from the herd for a couple of weeks while her lamb was small and vulnerable. Individually, they walked and walked and walked. Impala lambs are up on their feet in less than half an hour after birth. Despite having a small lamb on tentative, spindly legs, the females increased their home ranges to twice what they'd been before. Perhaps they were staying on the move in order to keep away from predators. If they considered Savimbi and I to be predators, then this strategy was certainly working. We were walking around ten kilometres a day up and down *kopje* after *kopje* just to keep up with them. The only way we knew that one had lambed was by her tracks. The mothers were always one step ahead of us. We'd be following a collared female for kilometres through remote parts of the farm, across areas she'd never used before, wondering how on earth we hadn't caught up with her, and then Savimbi would notice the small tracks in the dirt beside the mother's.

'You see, here this one is having a baby!' he would say jovially.

It was almost impossible to see the females during this time, which is testimony to their excellent mothering skills. When you have a vulnerable youngster in an environment lorded over by all manner of predators coming at you from every angle, including the sky, you have to be really street-wise to get your youngster past first base in life. Baby impala are like caviar to predators. They can't run very fast and they don't have horns to defend themselves. They're a lot better off than baby humans, who are totally dependent on their mothers, but the sad fact is that most juvenile impala don't make it. Although all of my radio-collared females gave birth, within two months almost half of their lambs were dead. It broke my heart, but it was perfectly normal for a prey animal living in the African bush to have a low rate of survival.

Mother-to-young ratios in the park suggested that only a quarter of black-faced impala actually made it to adulthood. Those that did were true survivors. It shed a new light on the way I looked at the adult black-faced impalas in the park. The odds were against them even making it to adulthood, and if they made it that far, the males still risked death from being gored by other males during the rut and they all still had to face up to the predators every night. The textbooks said that impala could live up to about twelve years old, but from what I'd seen, only a minority made it to the age that Grandma had.

Despite this, in Etosha the population was thriving. The two hundred black-faced impala that had been fortuitously translocated to the park thirty years ago by some far-sighted conservationists had built up to one thousand five hundred animals. But what were we going to do about the problem of hybridisation? How many black-faced impala were actually left in their historic range? Given that impala readily move through fences, did we need to protect the black-faced impala in Etosha from common impala on farms bordering the park? Lots of questions swirled in my mind. I'd been working on the subspecies for almost two years and it was time to start making a plan for their management.

I asked Werner from the institute if he would consider working with me to produce the first national management strategy for black-faced impala. To my surprise, he agreed. His boss, the director of research in the Ministry of Environment and Tourism, Dr Pauline Lindeque, fully supported us. I'd always admired Pauline, a strong woman at the top of a male-dominated field who had also retained a proud aura of femininity.

We decided to call a two-day workshop of selected government scientists and managers from all over the country to discuss the way forward. Among them were some of the biggest names in conservation in the country. Nothing happens fast in the Namibian government, but finally, in early April 2002, the experts gathered to discuss the future of my study species. I was one of only two women. The other, Dr Betsy Fox from Outjo, had worked as a scientist in Etosha for many years. While the men shook my hand cordially, grunting their acknowledgements, Betsy reached out and gave me a hug. It helped settled my nerves.

If my doctorate was going to do more than take up space on a dusty library shelf, the results had to be translated into management actions. Once my project was over, it would be up to these government scientists to enact measures for the conservation of the black-faced impala.

After the first day I walked out feeling disillusioned and disappointed. There was interest and input from the staff, but would this workshop actually make a difference? It was nothing but talk. Peter Erb, ever the stickler for procedure, asked why we were even bothering to write a management plan for black-faced impalas when we didn't know that they were genetically distinct from common impalas? I felt insulted on behalf of black-faced impalas that it would take a genetic marker to provide sufficient reason to conserve them. Is that what it had come to, that an animal had to prove itself as unique genetically to have a place on this earth? Scientifically, he had a valid point, but I knew this meant that any management action would be delayed until a genetics study had been carried out. I wanted action and I wanted it now! Patience has never been a virtue of mine. I remember selecting Saint Birgit as my confirmation saint because Mum had told me that I was in need of patience. Over a decade later, Saint Birgit wasn't helping me much.

As far as I was concerned, black-faced impalas were an important component of Namibia's arid-adapted biodiversity and were worthy of conservation on that basis alone. I told the gathering so, and a few heads in the room nodded. There was a glimmer of hope, but it wasn't enough. After two days Werner and I were able to write a reasonable management plan based on the workshop's discussions, but all actions would be on hold until we knew for sure that black-faced impala were genetically unique. I was running out of time.

We recruited a budding student from the University of Copenhagen, Eline Lorenzen, to undertake a genetics study. Efraim, my trusty tracker in Etosha, and I collected hundreds of vials of impala dung for her, from which she could derive cells for analysis from the mucus around the bolus. Shayne tirelessly shot at dozens of black-faced impalas in the park with a special compound bow. The bow shot a dart that collected a tiny piece of skin or hair when it hit the impala before bouncing out onto the ground to be collected. He reckoned it was the best fun he'd had in ages, and joked that it was probably the only chance he'd ever have to shoot an endangered black-faced impala. I gave Eline the eleven blood samples from the females we'd collared at Ongava. But it would be over a year before we had the result that the black-faced impala's DNA was indeed different from common impala, so the subspecies status was justified. Nothing happens quickly in Africa.

An unsettling sense of a phase of my life ending loomed as my date of departure grew closer. It hung over me like a dark cloud. At around the same time Tristan was offered a job running a nearby game reserve. The job was made for him. The timing was uncanny. In the room I called my office in the Etosha Ecological Institute, the computer file containing all of the data I'd collected during two years in Etosha mysteriously disappeared. Johan was perplexed as to how this terrible accident had happened, but I had him to thank for the fact that I'd backed it all up on CD. Somehow the loss of the file felt like an omen that it was time to move on.

I felt like part of the furniture in Etosha, more at home there than in Australia. There was so much I would miss about living there. The sonar ears of kudu, swivelling beside deep brown, intelligent eyes. The way the black-faced impalas bounded away with their white, fluffy tails flared, kicking up their hind legs as if to have the final say. The custard-coloured grasses and the way they danced in the wind, like a giant sea of grass. The vastness of the Etosha pan, prehistoric and glaring, a lunar landscape in the midst of a living, breathing universe of wildlife. The way that springboks pronked during the wet season, high on life, leaping into the air with all four feet off the ground, their backs arched and heads bowed. The blur of coffees and creams, stormy greys and bright whites, stripes and dots that made up the great herds of plains game. The way an elephant approached a waterhole, with his trunk raised in an S-shape to assess the smells; the way his wrinkles filled with moisture as his dexterous trunk sucked in the water urgently and poured it into his dry mouth with its nubile, hairy lips, his small eyes, dried and encrusted with dust, closed in ecstasy as the sweet liquid calmed his parched throat.

My friends in Etosha had made the place feel like home. What would life be like without dear old Nigel, with his characteristic sayings, such as, 'Rather burp and taste it than fart and waste it'? Would I ever have a friend like Shayne again? If it hadn't been for his and Birgit's regular invitations for a roast oryx dinner and their establishment of social Okaukuejo hockey games, would I actually still be sane? And as for the rest of the boys' club, it had taken me two years to gain a small level of respect and now I was leaving.

And then there was Tristan to consider. I hadn't wanted to pursue any kind of long-distance relationship, but that was what I was about to do. Did I have a future with this man outside this isolated environment? It was a confusing time, shadowed by doubts and fears. I tried to reassure myself that an ending is always the beginning of another adventure. But I wasn't looking forward to this one.

After a teary goodbye, I drove out of Etosha for what I thought would be the final time in August 2002. As I took in the hostile environment that I had grown to love, I knew that Etosha was branded into my life history; I would never be the same after my encounter with this place.

Just then I looked out to the right and saw a foraging herd of impalas near Ombika waterhole. I watched them for a second and then, as I turned back to face the road, a magnificent black-faced impala ram leapt over the bullbar of my car. I didn't even have time to slam on the brakes. Like a ghost, he seemed to glide through the air in slow motion, a flash of red-brown fur and lyre-shaped horns, before galloping off into the bush.

My heart almost stopped in fright at the thought that I might have hit him. Then I smiled. The impalas had always been smarter than I was. The ram was just giving me a final reminder of that. Maybe it was his way of saying goodbye.

A day later, high above Africa, the shifting, golden dunes of the Namib desert undulated and rolled below me, like waves on a sea of sand. The plane hovered low over the dunes, sprawling on forever. In the distance, a faint mist became visible on the horizon.

'The ocean,' Jen said, pointing ahead of us.

In my last week in Namibia before I flew back to Australia I joined my friend and fellow doctoral student, Jen Lalley, and Tristan's eighteen-year-old sister, Victoria (Pookie), for a week of girls-only quality time at the Skeleton Coast. This primeval landscape was Jen's study site. She was studying the lichens, the small, colourful plants that hold soil together in the desert and grow on rocks, with a view to gaining a better understanding of the human impact on this fragile desert ecosystem. Jen looked a little like an angel, with a halo of golden hair and deep blue eyes, but it was her innate strength that I had grown to admire. I could always depend on her for some straight-down-the-line girlie advice over a glass of dry white when I visited her home in Windhoek. Pookie too had angelic qualities, always able to fill a room with laughter and light with her exuberance for life. We had met through Tristan, but soon realised that we had huge amounts in common.

I looked out the window, awe-struck at the sight of the ocean, growing more excited by the moment. Shockingly sparkly, the vast cobalt-blue waters of the cold Atlantic formed a sharp contrast against the pastel harshness of the desert dunes. The pilot flew along the coast, dipping the aircraft low over a colony of seals. There were hundreds of them sunning themselves lazily, their glistening, rotund bodies awkward and cumbersome on the land. Some of them appeared to be sleeping, sprawled out, using each other's padded bodies as pillows.

Silent shipwrecks adorned the foreboding coastline, skeletons of a bygone era. Even if you survived the shipwreck and managed to find the coast, you were faced with the dismal prospect of a desert that stretched towards the horizon, into sands littered with skeletons. How many had bitten the dust here, I wondered.

A primitive airstrip appeared out of nowhere, and a gathering of large tents on decks, like small canvas houses. The pilot managed to plant the aircraft down, despite the heavy crosswind gusting in from the ocean. We were greeted by the burly figure of Chris Bakkas, the manager of the Skeleton Coast Camp. His long, fiery hair flew about his face under his Namibian bush hat as he grinned widely and roared in a deep voice, 'Hi, I'm Chris... Welcome to *hell*!'

As he reached out a massive hand to shake mine I noticed that half of his other arm was missing. He'd lost it to a crocodile, but it clearly wasn't the least impediment to this feisty fellow as he drove a Landie over the dunes like a pro.

Soon we were deep in the desert, surrounded for mile upon mile by sand and silence. There was not a tree or an animal in sight, and the only water seemed to be rather heavily salinated. That anyone at all could live in this harsh environment was hard to fathom. And yet, we were here to help Jen with her fieldwork on lichens. If there were plants, that meant there could be people and animals, which meant there had to be fresh water somewhere.

After a couple of days of counting and identifying lichens of every description, Jen took us to a nearby Himba village. She had promised to bring an old woman who lived there a copper bracelet for her arthritis. Jen's Land Rover chugged along the sandy tracks that snaked through the desert. Eventually, we came to the edge of a dry riverbed. Suddenly, there was life blooming in the desert. Large acacias and shepherds' trees lined the river, subsisting off the moisture they obtained from the water table below the surface. I observed the tiny, delicate lichens, resplendent in bright oranges, black and pastel greens. Elephant footprints meandered down and across the dry riverbed. A group of twelve desert-dwelling giraffes, tall and stately, with several youngsters, foraged on tall bushes, their delicate lips prodding through long, sharp thorns for their meal of leaves.

'How do the animals survive?' I asked, enraptured. 'I mean, there's no water around here, is there?'

'Fog,' Jen shouted above the rumble of the Landie. 'It comes in at night from the ocean and condenses on the vegetation, on rocks and even on the backs of beetles. Most animals get all the moisture they need through their food.'

Pookie and I nodded, inspired by the incredible diversity of life in this harsh place.

We'd been driving for an hour or so, tracing the path of the riverbed, when a person appeared out of nowhere. Clad in nothing more than a goatskin cloth, a meagre flap to preserve his modesty, he stood on the high bank of the riverbed staring down at us. He held a long wooden stick, like a staff, in his hand and was balancing on it, with the sole of his foot resting on the inside of his knee. A little further along, we encountered another man, also on his own, watching us. I noticed that the number of animals' trails had increased and then I saw the first sign of human habitation – a herd of patchwork-coloured goats.

'We're close,' Jen announced. 'See that hole in the river? Looks like someone's been digging there. The Himbas dig up the dry sand in the river until they hit water and that's where they get their water from – the people *and* the goats.'

I looked at the stagnant pool of mud which Jen was pointing at.

'Isn't that, you know, a bit unsanitary?' I exclaimed.

'Well, by our standards, sure it is. But these guys have lived like this for years. They're desert nomads. But you're right, it is unhygienic. This tribe has the highest rate of child mortality in the country. Something like one in three children dies.'

I shook my head. An infant mortality rate like that was incomprehensible to my modern Australian mindset.

As we rounded a bend in the course of the river, a trio of huts became visible. Long sticks had been bent into an igloo shape and then coated with cow dung to create simple homes. I noticed that there were plenty of gaps between the sticks for wind to blast through on cold desert nights. It was a harsh existence. Less than three metres in diameter, the conical huts were the epitome of the theory of minimalist living. Perched up on an open clearing, there wasn't a tree in sight around the village, a symptom of heavy overgrazing by goats.

Off to one side, a gathering of men had crouched down, pow-wowing among themselves. As we drove towards them, they all stood up together, like an army troop braced to defend themselves. Jen stopped the Land Rover a little way from the group and walked over to the hut. An old man stepped forward first. His face was wrinkled and hardened. A brown and red checked blanket was draped around his shoulders like a cape. The persistent wind was trying to blow it off his shoulders, but he held it tightly with one hand at his chest. It was clear from the man's proud demeanour that he was in charge. Jen shook his hand first, confirming that he was the chief.

Polite greetings were exchanged with handshakes between one and all, but none of the Himbas spoke English and only spoke a little Afrikaans and we didn't speak Ovahimba. Pookie translated what she could in Afrikaans.

Near the huts, three Himba women were sitting with a couple of small children. With their bodies painted in red ochre, copper bracelets around their arms and just a goatskin skirt around their waists, the bare-breasted women were a picture of simple beauty. The wind whipped the sand up onto the dirty goatskin on which they sat with their babies. Two small barefoot boys were playing nearby with a puppy, encouraging it to grab a stick in their hand, then hitting the puppy with it when it got close enough. The puppy kept yelping each time it got hit, then came back for more punishment.

One of the women sat cross-legged with a toddler in her lap. While still occupying the toddler with one arm, she lifted a floppy breast to another baby's mouth and forcefully shoved the nipple in. The toddler stood up, wobbled a little on her chunky legs and then proceeded to fall over in the dirt, landing on her chubby, brown buttocks.

She sat there for a moment, contemplating, the way small children do, and a strange, confused look came over her face. Suddenly, a creamy green liquid oozed out from her robust buttocks. Her mother shouted something, heaved the smaller baby off her breast and placed her on the blanket beside her. She grabbed the little girl's arm roughly, continuing to verbally reprimand her. Without yanking off the little cloth the toddler wore around her waist, she used another blanket to wipe the green mess off her buttocks. Now the little girl began to whimper.

Grabbing the first aid kit from the back seat of the car, I discovered a packet of Imodium under the many boxes and bandages in there. *For treatment of diarrhoea, adults one tablet only*, the type on the box read. A full tablet would probably clog her up for good, Jen said, but the likelihood of these women getting any medical help was unlikely. This community was a five-hour drive across the desert to the nearest clinic in the village of Sesfontein. We'd have to find a way to lower the dose. The tablets were in capsule form, so I cracked open the capsule and poured some of the white powder into a plastic cup. After mixing it with water, Pookie tried to explain to the mother what the drink was for. I offered it to the mother, pointing at the little girl and indicating that she should give it to her to drink.

But try as she did, the woman could not make her daughter drink. Her stomach bulged, a sure sign of malnutrition, and I noticed now how weak she was. The little girl had simply curled up on the blanket and closed her eyes. She was probably already severely dehydrated, if the state of her diarrhoea was any indication. Suddenly, the mother made her sit up and pushed her over to me, indicating that I should help. Forcing the little girl's head back while her mother held her still, I poured as much of the rotten-tasting drink into her mouth as I could. Half of it dribbled out, and only a little seemed to go down. The child was by now quite horrified by the strange white woman who was forcing her to drink something that tasted disgusting.

We had to think of another plan. Chocolate: the remedy for all ailments, physical, emotional and spiritual! Jen had some in the car. After mashing some of the diarrhoea powder into the soft chocolate I put it in the child's mouth and let her suck on it for a few seconds, so she could realise it tasted better than the disgusting water mixture. Her eyes widened as she sucked the sweet chocolate with fascinated curiosity, before it was replaced with a look of sheer terror when she remembered who was feeding it to her.

Thirty percent infant mortality. It was no wonder. This was a place where dehydration killed people and all it could take was a little dose of diarrhoea and your baby was dead. There was more than one reason why it was called the Skeleton Coast. We mixed a packet of rehydration salts into a bottle of mineral water and instructed the mother to give her daughter sips of this solution throughout the day.

With the conversation dominated by mixed feelings about our sobering experience at the Himba village, that evening we drove out on top of a sand dune to un-sober ourselves over sundowners. The three of us sat on the sand, rugged up in beanies, jeans and jumpers to keep out the famous Skeleton Coast wind, and toasted life.

As the white orb of the sun sank behind one dune, we turned to face the opposite direction. There, the full moon rose, as red as I'd ever seen it, changing the mood of the land as it swelled in the sky.

'Blood on the moon,' Pookie commented thoughtfully, echoing my own mind.

The dense Atlantic fog started to roll in, tumbling like billowing smoke inland, filling the desert with moisture. It was getting cold. It was time to go.

AN ENDING AND A BEGINNING

A gust of warm breath blew into my face. It smelt like composting vegetables. Natural and wild. In what appeared to be a peaceful sleep, lying on his side, the elephant bull breathed in deeply, filling his giant lungs with air. His exhalation, slow and deep, sounded as though it came from the sacrosanct depths of the ocean. A small stick perched in the hairy, malleable tip of his snorkel-like trunk held open the small airway, allowing air to flow in and out.

'Your breath smells,' someone said to the elephant, trying to make a joke.

A buzz of Earthwatch volunteers from America and England swarmed around the bull. One of them was throwing a bucket of water over the elephant's wrinkled body, cooling it down as it lay in a drug-induced state of blissful ignorance. Keith Leggett bore a worried frown. This operation was his responsibility. It was a whole lot more dangerous than the work I'd helped him with in Kaokoland when I'd just started in Namibia as a fledgling doctoral student.

AN ENDING AND A BEGINNING

It was now 2004, I had my doctorate and I had been back in Namibia for over a year. Somehow, mostly through a sheer stubborn dedication to getting my data analysed and written up as quickly as possible, I'd written my PhD in a nonstop nine months. I became a nerd of the highest degree and disappeared into the vortex of statistics, scientific literature and long, often mind-boggling days at the computer. Initially I felt as though the mountain of work was so large and daunting that I would never make it to the top. Away from Etosha and the wide open spaces of nature, I felt lost and out of place. The one thing that kept me going was the knowledge that the sooner I started the sooner I would finish. The reward at the finish line was a return to the land – and the man – I loved. On 30 June 2003, my twenty-sixth birthday, I boarded the plane back to Namibia, having handed in my thesis two days before. It felt like there wasn't a moment to lose. After so long away, I thought I would literally fall apart if I left even a day later.

The year that followed wasn't always easy as I struggled to find work, battled to get funding for a research project and generally adapted to life in the strange stage that often follows a PhD, after the dream that had consumed my life for so long was finished. I struggled to get a work permit and faced setback after setback where funding and research opportunities were concerned. The ongoing disappointments and the struggle to find enough money to pay the rent each month took their toll on my relationship with Tristan as well as on my enthusiasm for wildlife conservation in Africa. But at times like this, when I was crouched beside an enormous desert-dwelling elephant as he breathed in and out on the sands, I remembered why Africa had such a hold on me. All the answers were in nature. It was simply a matter of seeing them.

Keith and wildlife vet HO Reuter were scooping sand away from underneath the bull's neck and pushing through a massive satellite collar. The collar was about three metres long and made of heavy machine belting. At one end a battery pack and satellite transmitter were encased in solid dental acrylic and weighted with concrete. This was the part of the collar that would relay GPS coordinates via satellite to Keith's computer for the next year. Unlike the radio collars that I'd used on the black-faced impala, Keith was using far more sophisticated technology that meant he wouldn't even need to leave his office to track his elephants. When I tried to lift the collar, I realised how heavy it was, but to the bull, who weighed over a tonne, it was as light as a feather. Andy, a gregarious English mate of Keith's, drove in the screws to fasten the collar around the elephant's neck. It was fixed with stainless-steel bolts. The collar had to be strong enough to survive the brutal day-to-day life of being attached to an elephant bull in the desert.

The bull was breathing steadily as HO prepared the M99 antidote. Keith was taking the final body measurements, noting the length of the bull's dirty tusks. I examined the bull's feet. Elephant feet have always amazed me. How such a large creature can be so silent in the bush is nothing short of incredible. The bull's padded soles looked like old concrete, dusty and wrinkled, his toenails stained yellow and hard as a rock. These were feet that had done some walking. One of Keith's study animals had walked over six hundred kilometres in just a few days.

The elephants of Namibia's arid lands aren't your average, run-of-the-mill elephants. They're desert-dwellers. The way they survive is by walking vast distances from one patchy resource to the next, following the dry riverbeds, trusting their innate knowledge of where the water lies, digging it up where they have to, and recalling precious seasonal memories of where they can find ana and acacia trees overflowing with nutritious pods.

AN ENDING AND A BEGINNING

'The other unique thing about desert-dwelling elephants,' Keith explained to me, scratching his day-old white stubble, 'is that they don't have matriarchs. They form loose aggregations and small herds in order to survive. It's an arid zone adaptation, and a matter of survival.'

They can't afford to form large, cohesive herds led by a dominant female as elephants do in more lush parts of Africa, because there isn't enough food to go around in the desert. Instead they are nomads, hardened survivors in a moonscape of undulating sands. It is more or less a case of every elephant for himself.

I touched the wrinkled, grey skin of the bull's chest. It was hard and rough, folded into deep wrinkles and sprouting bristly hairs as tough as metal wire. In the background Tristan stood with his .416 Bruno rifle, keeping an eye on two other bulls that had been with the tranquillised bull when HO had shot him with a dart of M99 earlier. If they came too close to the unconscious elephant and the gathering of people, Keith had told Tristan to shoot into the air to ward them off.

My nerves, like Keith's, were on edge. Elephants are very smart. To these bulls, their friend probably looked dead and surrounded by a tribe of unpredictable humans. I wouldn't have put it past them to give us a run for our money. In fact, a charge would have been justified. Luckily, the bulls lost interest and moved on. But no one could let their guard down. A stately desert-dwelling giraffe peered with apparent fascination at the scene from behind a camel-thorn tree where he was browsing. Obviously this wasn't the sort of thing the giraffe saw every day out here in the desert bordering Namibia's Skeleton Coast.

After twenty busy minutes, with the massive collar fastened and Keith satisfied that all of the body measurements had been taken, HO injected the antidote. Everyone piled into the Hiluxes and pulled away through the thick sand. I was driving Tristan's Hilux and prayed that I wouldn't get bogged in the dry riverbed's loose sand with the back loaded with foreign, paying volunteers and a number of annoyed elephant bulls in the near vicinity. I stopped a little way back and we watched the bull come to his senses. He rolled around like a goldfish flailing on land, trying to lift his heavy head first, and then attempting to move his trunk, which obviously wasn't doing what his brain was telling it to. Ungainly and awkward, he used the momentum of his own body to get up onto his feet, reversed a bit in a drunken stumble, and then stood still, trying to find his wobbly footing. His giant penis suddenly slithered out of its sheath in a huge erection almost the size of his leg. The bull defecated, leaving a steaming pile where he stood, and then began to lumber away from the smell of humans as quickly as he was able, his massive erection bumping off each back leg as he walked.

We all breathed a sigh of relief that the operation had gone so well. Keith had several elephants that he wanted to either collar or de-collar in this area, so the job wasn't over yet. But for today, our work was done and we all drove back to the community campsite for lunch and a midday siesta.

Set on the wide, dry Hoarusib River, the campsite was run by the Himba community at the nearby village of Purros. It was not far from the coast and the cold breath of the Atlantic Ocean kept the desert dunes relatively cool, despite the relentless sun that warmed the top layer of sand. Tall ana trees and camel-thorn acacias towered over the shady campsite and along the length of the dry river's banks like welcoming friends. Humans and elephants alike were drawn to these shaded, sacred places of rest. Not far from the campsite Tristan and I had seen rare desert-dwelling lions on an oryx kill, as well as springboks grazing and elephants browsing along its vegetated banks.

AN ENDING AND A BEGINNING

Tristan and I raked away fallen branches, thorns and piles of dried elephant dung from beneath an ancient ana tree to create our own campsite away from the crowd. We laid down a canvas tarp on the sand and deposited our swags on top of it. A thick, vibrant patchwork quilt that Mum had made for me topped off our desert bedroom's furniture. We both collapsed on the welcoming bed and Thau, Tristan's beloved dog, lay on the sand at our feet. It was a hard life.

About an hour later I was woken by a jab from Tristan's elbow. I opened my eyes, still half asleep, to see the blur of a young elephant bull pulling pods and leaves from an acacia tree about fifty metres away. My vision quickly came into focus. There was a meagre stretch of sand between him and us. But he didn't seem too concerned about us. We watched him in silence for five minutes, enjoying the magical sight at close range from our beds.

Very quietly, smiling, I said to Tristan, 'I'm getting into the car if he starts coming our way.'

'Just stay still,' Tristan replied. 'He won't come.'

A minute later the bull decided that he'd had enough of the tree he was feeding on and began to head for the nearest available ana tree. It just happened to be the one we were lying under. I looked up to the canopy. The tree was laden with twisted, nutritious pods. Ana pods are a food so treasured by elephants and giraffes in the desert that the browse line in this tree was over fifteen metres high, up to where only the tallest bulls could feed. Elephants passing by had balanced on their back legs and reached up with dexterous trunks to pull the pods down. Now this bull had his eyes on some within reaching distance. Led by his stomach, a typical male, he only had one thing on his mind: food.

Each of his steps seemed to occur in slow motion, giant steps over a short distance. In seconds he would tower over us and it would be too late to get away. I had a vision of Mum's patchwork quilt being used to wrap our flattened bodies in for burial.

'I'm getting in the car,' I announced, readying myself to run.

Thau was still sound asleep at our feet, snoring obliviously.

'Just stay still,' Tristan said again, but less confidently this time.

Then suddenly he commanded sharply, 'Thau! Get in the car.'

The snoozing dog responded instantly, looked up to see a tonne of desert elephant heading his way, put his tail between his legs and leapt into the back of the Hilux as if poked with an electric prod. I followed with equal fervour. In the back seat, next to Thau, I crouched trembling with a mixture of excitement and genuine fear. If it hadn't been so terrifying it would have been funny.

Tristan grabbed his rifle. He stood beside the ana tree and waved his hands in the air to let the bull know we were there, but the elephant kept coming. There was no question that he wanted to get to the tree that was our camping place. We were a troublesome interference in his feeding regime. My eyes widening by the second, I watched as the bull came around the other side of the tree to where Tristan stood. Tristan continued to wave his arms and rifle around in the air, shouting at him, 'Tsk! Tsk!'

The bull was now five metres from us and my neck was kinked with the effort of looking up at him. He loomed over us, making us feel like mice in the presence of a giant. He was angry. He flapped his ears, making a sharp sound like a mat being hit on a cement floor. Dust billowed off them. He raised the front of his body up, enlarging himself even more, trying to intimidate Tristan who was still standing up to him, not aiming his rifle but using it to tell the bull to bugger off with rapid movements and noises.

'Tsk! Tsk!' Tristan shouted vehemently.

Thau whimpered quietly in my arms. Face to face, eye to eye, the young man and the young bull, not so different from each

AN ENDING AND A BEGINNING

other, stood their ground. The bull didn't want to give up his tree. The man didn't want to give up his campsite.

This went on for an excruciatingly tense twenty seconds before the bull spun around and walked away with a gait that said, 'You're not worth it, mate.'

Tristan, to our relief, had won this stand-off. I realised that I was shivering and it wasn't from the cold. Thau jumped out of the car and ran up to Tristan with unbridled joy. I looked over to Keith's campsite where a gathering of volunteers was standing watching.

'That was a bit close!' one of the girls called out, grinning from ear to ear.

It was the understatement of the century. The young bull had had every right to be annoyed with us. We were in his territory, after all. Tristan had shown great courage by standing up to the bull, but so had the bull in facing up to Tristan. Humans had been responsible for the persecution of his kind throughout much of this region, and elephants have long memories. Ironically, it was now in human hands that the elephants' future lay.

But in some places there wasn't enough space for everyone. With growing populations of both elephants and humans and, concurrently, available habitat for both species reducing in area, conflicts with humans dot elephant migration paths in much of Africa today. In Namibia people and elephants are often competitors in a harsh desert land where resources for traditional agriculture are generally poor. Many rural people live below the poverty line and depend on the meagre crops and water installations that elephants often destroy. To ask a man who doesn't have enough to feed his family to conserve the animal that destroys his livelihood is contradictory to say the least. Then again, most people would agree that conserving viable populations of elephants in Africa is absolutely crucial. In Namibia, as in much of southern Africa, a possible solution is being heralded to the conflicts between rural people and elephants. It's known as Community-Based Natural Resource Management.

Being largely a desert land, most of Namibia isn't ideal for traditional farming. However, one thing that Namibia does have is some of the most breathtaking landscapes in the world. The grandeur of the desert, the magnificence of its canyons, mountains and coastline and the incredible peoples, as well as the local animals and plants can't be compared with anywhere else in the continent. Namibia is a jewel.

As a result of its valuable natural assets, efforts are now being made to empower local communities to conserve wildlife, while simultaneously improving the quality of their own lives. Elephants are paying their way through community-based joint ventures. Indigenous groups throughout the country, like the Himba community at Purros, realise that they can make money from wildlife-based tourism and trophy hunting. Non-government organisations such as the Worldwide Fund for Nature, and environmentally-minded companies like Wilderness Safaris, are helping to develop skills so that ultimately rural people perceive elephants as a benefit, generating income, jobs and self-sufficiency, rather than a problem to be dealt with using short-term treatments.

I had always been interested in the people side of wildlife conservation. My experiences had shown me that dumping a heap of foreign aid on battling rural Africans was only a short-term solution to their problems. Their magnificent wildlife and ecosystems are their assets, so why shouldn't they be able to use them to find a way out of the black hole of poverty and powerlessness? Rather than giving rural people handouts of clothes and food, doesn't it make better sense to help them with skills training so that they can make something of their lives? Rather than giving them western clothes, why not foster a sense of pride in their own extraordinary traditional cultures?

Such thoughts were spinning through my head when Dave van Smeerdijk, the charismatic Aussie managing director of Wilderness Safaris Namibia, offered me a job helping to run the ecotourism company's environmental division. Initially my job was to supervise a recent science graduate, Basilia Shivute, a shy Owambo girl who was a few years younger than me. However, I soon realised that Basilia had a lot to teach me. She was gutsy, determined and took great pride in her job, despite being the only young black woman at the upper management level of the company in Namibia. The two of us never failed to attract stares from tourists when we ate together at the camps, as if a black and white woman sitting together at a table was a strange sight, but to us the only strange thing was their staring.

Basilia, who was often shy around other people, would regale me with stories over dinner, and for a little person, man could this girl eat, especially if a game steak was on offer! She told me that she'd never been very good at working in the fields as a child growing up in rural Owamboland, being short and petite. Her mother used to keep her inside and tell her that if she didn't want to become a traditional Owambo wife working in the fields for the rest of her life then she'd better work hard at school and go to university so she could get a good job and be self-sufficient. That Basilia had achieved this at the age of twenty-four made her something of a role model for Namibian women of her generation.

Together we were a two-woman team responsible for making sure that Wilderness Safaris' twelve camps were run environmentally sustainably. The camps were located in some of the most pristine environments in the country. We either flew to the camps in light aircraft or drove, sometimes for days at a time, through some of the most remote and primitive places I have ever seen. On one noteworthy occasion we had to trust our instincts to find our way through the Hartmann's mountains in Kaokoland with a sketchy hand-drawn map, in an area so remote that we didn't see another vehicle on the road for a couple of days of driving.

Wilderness Safaris established most of their camps in partnership with rural communities in Namibia with the goal of training members of the community in tourism, employing them in the camp and providing the entire conservancy with a percentage of the income derived from paying tourists. The longer I worked for Wilderness Safaris and the more I talked to Basilia about the issues facing rural people in Namibia, the more I saw great benefits for wildlife and communities if companies put their money where their mouth was. There were still lots of wrinkles to be ironed out to maximise the benefits to the communities, but on the whole we seemed to be going in the right direction – for people and wildlife. I was glad to be a part of it.

I'd been working in Wilderness Safaris for about six months when I visited Bushmanland in north-eastern Namibia. This is the home of the San people or Bushmen, locally known as the Ju/'hoansi people (in their pronunciation the symbol '/' represents a click of the tongue). Separated from the rest of Namibia by a straight, wide gravel road that stretches for three hours from the nearest town, Grootfontein, the capital of Bushmanland is a shanty by the name of Tsumkwe. It consists of two shops, a community hall, a curio shop, a few shebeens and a couple of government houses. Simple huts made of grass, sticks, mud and sometimes corrugated iron house a few families on the fringes, where donkeys and cows wander the dusty landscape. A wide gravel road snakes in an L-shape through the heart of Tsumkwe. In thirty seconds, you can drive at a moderate pace from one end of the town to the other.

AN ENDING AND A BEGINNING

The government warden for this area, Dries Alberts, warned me: 'Be careful you don't drive past it. If you see a chicken, you've just driven past Tsumkwe.'

I was visiting Bushmanland to talk to the community about possible wildlife research projects in their conservancy, Nyae Nyae. My future was uncertain to say the least. My part-time work for Wilderness Safaris was enough to pay the bills, but I wasn't sure if it was a long-term career prospect. And something about Bushmanland captured me. I felt as if I was under a magic spell. I knew I would come back, but I had no idea how.

There is something very special about the Bushmen. For a start, they're tiny. The grand old chief of Bushmanland, the Honourable Chief Bobo, stretched out a wrinkled hand to me in welcome to his land. He was shorter than I was. His almost Asiatic eyes crinkled as he smiled, his face glowing with genuine warmth, wisdom and peace. His old clothes hung off his gaunt frame. Only his belly and bottom were well rounded, as seems to be common to the Bushmen. He spoke no English, only Afrikaans and his own language, spoken in clicks. Here was the chief of an African kingdom, a politically appointed traditional leader of a tribe of people, and yet he was possibly the most humble person I have ever met. As he explained to me through a translator what his people needed help with to better manage their wildlife, he laughed and smiled. Life was indeed a merry thing to him and his people. He reminded me of the Dalai Lama.

The irony was that, like the Tibetians, things hadn't been too great for the Bushmen for quite a while. They had been forced into an area only fourteen percent the size of that which they'd used before, an area designated Bushmanland by the previous South African government. Alcoholism and domestic violence, along with western ways, clothes and convenience foods, had pervaded the social environment, eating away at its delicate fabric.

Fewer and fewer of the younger generation were learning the old traditions. The old men couldn't speak English and it was still uncommon for Bushmen children to be schooled beyond Grade Four, many didn't go to school at all. The younger generation was battling to survive in modern-day Namibia as their culture drifted away with the elders. Most didn't have enough education to compete with other ethnic groups in the cities. The village children were afraid to go into Tsumkwe where there was a boarding school, let alone a place the size of Windhoek.

In the past the Bushmen had been shot at like wild animals and treated as the lowest of the low. The word San itself means thief, which is how they were once seen. Other tribes perceived them as primitive and compared them to animals, yet this was a simple, gentle people who smiled readily and welcomed strangers.

When it comes to conserving wildlife, the Bushmen have always been the unrecognised leaders in the field. They have lived in harmony with animals pretty much forever, treating the environment with the respect it deserves. As hunter-gatherer nomads, they killed only what they needed in order to survive and they have never been pastoralists. So this was the last place I expected to hear of a problem with human–elephant conflicts. Those indigenous groups with crops that elephants raided and destroyed had reason to be antagonistic towards elephants. The Bushmen, on the other hand, lived in harmony with the natural world. They were afraid of elephants because they destroyed the water installations and trees that they depended on for food and water. A few years before, a woman had been killed by an elephant while she was collecting bush foods. Despite this they valued elephants because they were an integral part of nature, and because they represented income and employment through tourism and trophy hunting.

The old chief explained in his clicking tongue that he needed help. One of the conservancy men translated for us. In the dry season, large concentrations of elephants migrated into Bushmanland, probably from Namibia's Caprivi area and Botswana's Okavango Delta. Bushmanland was on the border of the largest elephant population in Africa, Botswana, so it wasn't a problem that was going to go away by itself. Desperate with thirst, the elephants pulled up water pipes and destroyed the installations that provided water not only for them but also for the Bushmen and the other animals. The old chief explained that his people wanted the elephants to stay, but they had to find a way to stop them destroying their fences and water installations.

Chief Bobo's dilemma touched my heart. The more time I spent with the people actually living with wildlife in Africa, the less I felt that the academic scientific path I'd been following, with its rigorous conformation to man-made rules, could provide the answers. Science had its place, but more and more I felt it limiting my creative energies.

At twenty-six I'd achieved my dream to be a wildlife researcher in Africa. It had been a spiritual, emotional and intellectual pilgrimage that had caused me to grow from a teenage girl into a young Australian woman whose heart beat to an African drum. But now that I'd achieved that, what was my next dream? It was only through sheer drive and determination that I'd made it to the end of my doctorate, it certainly wasn't through any natural ability as a scientist. I didn't have the perfectionism and attention to detail required to be a meticulous academic and I knew that I wouldn't be fulfilled by academia.

After all we'd been through, to throw a spanner in the works, I was beginning to wonder if Tristan and I were going to make it. With just enough money to survive and not enough work to keep me stimulated, I started writing. It was an escape from the haziness of my reality and took the focus off the fact that I had absolutely no idea where I was going next, whether it would be in Africa, with Tristan, or as a scientist. I was on the verge of a career change, although I didn't know what it was.

Through the fog of indecision, I began to see that, if nothing else, I had a story to tell. Africa seemed to be urging me to use my experiences to help its people and wildlife. The old chief of Bushmanland's gappy smile in a face of wrinkles, his skin gnarled and weathered like the ancient baobab trees of his country, mingled in my memory with images of the jubilant children of Humani School, Ipheas changing the tyre of the Blue Beast with a wide grin on his face, Jessie the cheetah stalking her first prey, Karen calling to the hippos, Savimbi tracking black-faced impala through the bush and the old bull elephant being collared in the desert sands.

My journey in Africa wasn't over, not by a long shot. In the next chapter of my life I knew that I would have to listen to my head but only trust what was in my heart. It's not always an easy ride, but dreams do come true. My life was living proof of that. Even though I didn't know what was next for me, I knew that my African journey had only just begun. If I'd made my dreams come true once, then I could do it again. To succeed, I reckon, you don't necessarily need much in the way of natural talent, but you do need the determination and perseverance of a dung beetle or a honey badger. Most of all, I have learnt that even when life feels out of control, you must believe in yourself. And then you will have magic in your life, leaving a trail of dreams and dust in your footsteps.

ACKNOWLEDGEMENTS

This book would never have happened if it were not for the amazing support network of family and friends – all over the world – who gave me the gumph to keep stepping into the unknown. My parents, Allan and Rhonda, my sister, Kek, brother, Davo and beloved niece, Ella – without your belief in me (even if it was under sufferance, Mum!) I would never have made Africa my home. I am totally indebted to Roger, Anne, Sarah and the Whittall throng, Karen and Jean Paolillo and so many other dear friends in Zimbabwe – I only hope that the Zim we all knew and loved when I was 'Scary's Slave' will return to its former charming self. Special friends who kept me going when the boys' club got too much – Sal, Shelby, Kise, Jen, Pooks, Naomi, Kel and so many other members of the cool chicks club – thank God for female mates when the going gets tough! The staff – my friends – in Etosha and at Ongava – you made my time there so much more than a PhD. My superb trackers – Ipheas, Efraim and Savimbi – thanks for not letting me get eaten or stood on. My supervisors, Anne Goldizen and Peter Jarman never gave up on me, even when they realised that I was doing a PhD for all the wrong reasons! Finally, to my publisher, Bernadette Foley, who saw something special in my story and convinced me that it might make a good read, and to Siobhan Gooley and Julia Stiles for their fabulous and humane, cross-continental editing of the drafts.

Africa has given me more than I can ever repay – I just hope that in this lifetime I can give a little back.

Dr Tammie Matson was born in Townsville, Queensland, in 1977. On her first trip to Zimbabwe with her father at the age of fifteen she fell in love with the country and decided she had to come back. She changed her degree from Law to Environmental Science so she could study there. After gaining her degree in 1999 she was offered a scholarship to do a PhD in Zoology which she completed in 2003. She has been living in Namibia since 2000, based initially at Etosha National Park and subsequently in the capital, Windhoek. She currently works as an environmental consultant in Namibia, for organisations including Wilderness Safaris, Save the Rhino Trust and the Namibian Professional Hunters Association, as well as running a research project on human–elephant conflicts at the request of the Chief of the Bushmen.